Florence Nightingale at Home

"This is the most important study of Florence Nightingale's life and work for a generation. Using newly available sources, the authors set her firmly within the political, cultural and spiritual life of Victorian Britain as someone who both transcended and reflected the roles traditionally available to women of her class."

—Professor Robert Dingwall, *Nottingham Trent University, UK*

"This brilliant book deftly uses the device of 'home' to interrogate the life and achievements of Florence Nightingale. By doing so it brings a new interpretive lens on the role that 'home' played in her reform efforts and breathes new life into this ever fascinating nurse and cultural icon."

—Professor Anne-Marie Rafferty CBE, *Royal College of Nursing, UK*

"*Florence Nightingale At Home* is a remarkable achievement, and the authors are to be congratulated on so effectively locating Nightingale in the context of her time. This book is not only an invaluable source for historians of women, nursing, medicine, public health and Victorian Britain; it is also an elegant and absorbing 'read' for a more general audience."

—Professor Christine Hallett, *University of Huddersfield, UK*

"This book is not only scholarly and accurate, but has excellent visuals and is wonderfully readable. The authors use the 'home' theme to present material from the Crimean War and press coverage of it, her faith and the pioneering research she did from her own home post-Crimea."

—Professor Lynn McDonald, *University of Guelph, Canada*

"A fascinating insight into some of the most personal yet valuable areas of Nightingale's life, this book will be a welcome addition to the library of any Nightingale enthusiast. The unique and timely focus of this book makes for a compelling read. It is a book marked by passion and dedication."

—David Green, *Florence Nightingale Museum, UK*

"This highly original book is not a typical biography. It provides a fascinating insight and a renewed understanding of Florence Nightingale through her varying experiences of home. The authors effectively demonstrate how changing ideas of home continually shaped Nightingale's attitudes and work, both in Britain and abroad."

—Dr. Vicky Holmes, *Queen Mary University of London, UK*

Paul Crawford · Anna Greenwood ·
Richard Bates · Jonathan Memel

Florence Nightingale at Home

Paul Crawford
School of Health Sciences
University of Nottingham
Nottingham, UK

Anna Greenwood
Department of History
University of Nottingham
Nottingham, UK

Richard Bates
Department of History
University of Nottingham
Nottingham, UK

Jonathan Memel
Department of English
Bishop Grosseteste University
Lincoln, UK

ISBN 978-3-030-46533-9 ISBN 978-3-030-46534-6 (eBook)
https://doi.org/10.1007/978-3-030-46534-6

Cover credit: courtesy of Claydon House Trust. Design by eStudio Calamar

This Palgrave Macmillan imprint is published by the registered company Springer Nature Switzerland AG
The registered company address is: Gewerbestrasse 11, 6330 Cham, Switzerland

To the victims of the coronavirus pandemic, and to all those who risk their lives to save others

Acknowledgements

We are grateful to the Arts and Humanities Research Council for funding the research project that informed this book (grant number: AH/R00014X/1). We wish to thank Professor Lynn McDonald for making searchable files of her sixteen-volume *Collected Works of Florence Nightingale* available to our research team; Sir Edmund Verney, Nicholas Verney, and Sue Baxter at the Claydon House Trust for providing substantial access to their archives; David Green and Chloe Wong at the Florence Nightingale Museum for facilitating access to research material; and Sarah Chubb of the Derbyshire Record Office and Stephanie Davies of Lotherton Hall for assisting with archival research. Thanks also to our formal and supporting project partner organisations, the Florence Nightingale Foundation, British Library, University Hospitals of Derby and Burton NHS Foundation Trust, Derwent Valley Mills World Heritage Site, Derbyshire Record Office, Florence Nightingale Museum, Royal College of Nursing, Southwell Workhouse, Queen's Nursing Institute, Royal Literary Fund, National Portrait Gallery, Friends of Aqueduct Cottage, Derby Museums, Lotherton Hall, Wellcome Trust, and John Smedley for their unwavering support; to our Advisory Board members, John Rivers CBE DL, Professor Anne Marie Rafferty CBE, Professor Chris Wrigley, Professor Brian Brown, Professor Susan Hogan, and Peter Kay, owner of Lea Hurst, for their ongoing encouragement; and finally to our distinguished guest bloggers and citizen researchers who joined us along the way. Thanks to the speakers and attendees at our project workshops, 'Locating Health' and 'The Home in Modern History and Culture', especially Professor Christine Hallett and Professor Jane Hamlett for their keynote lectures. Thanks also to Tim Clark for copy editing; to Marian Aird for providing the index; to Jamie O. Crawford for his early reading and editorial suggestions; and to Dr. Vicky Holmes, Frances Cadd, Mathilde Vialard, James Burrows, Dr. Ewa Szypula, and Peter Bates for reading and commenting on parts of the manuscript. Special thanks go to our dear colleague Hayley Cotterill, senior archivist at the University of Nottingham, and to the whole Manuscripts and Special Collections team, who enabled this research to be featured in the exhibition 'Nightingale Comes Home' at Nottingham's Lakeside Arts Centre.

We are grateful to the following organisations for permission to use images included in this volume: Claydon House Trust, an independent charitable trust caring for the archives of the Verney and Nightingale families at Claydon House; Wellcome Collection; National Portrait Gallery; Victoria and Albert Museum; Minneapolis Institute of Arts; and Fliedner Kulturstiftung, Kaiserswerth. All other images are either used under a Creative Commons licence or their copyright has expired.

Contents

About the Authors

Paul Crawford is Professor of Health Humanities at the University of Nottingham. He pioneered the field of health humanities globally, and has authored or edited 13 books. His new book, *Cabin Fever: Surviving Lockdown in the Coronavirus Pandemic* will be published by Emerald Press in 2021. He has been a Registered Nurse since 1989, and he is Fellow of the Royal Society of Arts, the Academy of Social Sciences, and the Royal Society for Public Health.

Anna Greenwood is Associate Professor in Medical History at the University of Nottingham. She has authored several books and numerous research articles, is a commissioning editor for the *Gender and the Global Health Humanities* series (Emerald Press), and is a Fellow of the Royal Historical Society.

Richard Bates is Research Fellow in History at the University of Nottingham. His research focuses on social history, gender, and medicine in nineteenth- and twentieth-century Britain and France. He has published work in *Journal of Political Ideologies*, *Journal of Medical Humanities* and *French History*.

Jonathan Memel is Lecturer in English Literature at Bishop Grosseteste University. He researches the relationship between literature, health, and education in nineteenth-century Britain and has published work in *History* *of Education*, *Neo-Victorian Studies*, and *Modern Language Review*. He sits on the Executive Committee of the British Association for Victorian Studies and edits their tri-annual newsletter.

Outline Timeline

1803	Derbyshire lead-mining magnate Peter Nightingale leaves an entailed fortune of £100,000 to William Shore, aged nine, from a Sheffield banking family.
1815	William Shore assumes the inheritance and takes the Nightingale name.
1818	William Nightingale and Frances Smith marry in London. They set off on a European tour lasting three years, most of which they spend in Italy.
1819	Birth of Florence's older sister Parthenope Nightingale, in Naples.
1820	Birth of Florence Nightingale on 12 May, in Florence.
1821	The Nightingales return to England and settle in Lea, Derbyshire.
1826	The family moves to Embley, Hampshire, an estate purchased for c. £125,000 in 1825. Thereafter, they spend each autumn/winter at Embley, the spring in London, and summer at Lea Hurst, Derbyshire.
1827–31	The Nightingale girls are educated by a governess, Sara Christie. From 1831, William Nightingale takes over their education.
1830s	Nightingale and her female relatives practise charitable health visiting to poor communities around their Derbyshire and Hampshire estates.

1837	Aged sixteen, Nightingale experiences a religious 'call to service' during an influenza outbreak in Hampshire.
1837–39	The Nightingales undertake a European tour, principally of France, Switzerland, and Italy, during extensive refurbishments to Embley.
1845	Nightingale's parents refuse her permission to nurse at Salisbury Hospital.
1846	Nightingale reads an annual report of the Kaiserswerth Deaconesses' Institution and visits the German hospital in Dalston, London.
1847–48	Nightingale in Rome with the Bracebridges. Meets Sidney Herbert.
1849	Richard Monckton Milnes proposes to Nightingale; she refuses.
1849–50	Nightingale tours Egypt and Greece with the Bracebridges, returning via Central Europe and spending two weeks at Kaiserswerth.
1851	Nightingale returns to Kaiserswerth, staying for three months.
1851–52	Writes 'Cassandra'.
1852	Nightingale visits Ireland. Parthenope suffers a nervous breakdown.
1853–54	Runs the Establishment for Gentlewomen During Illness in London.
1854	Britain and France join the Crimean War. In October, Nightingale departs for Scutari with thirty-eight nurses. She arrives in November and takes charge of the nursing operation at two military hospitals. Nightingale works with *The Times* to appeal to the public to send supplies.
1855	In February, the iconic depictions of Nightingale with her lamp appear in *The Times* and *The Illustrated London News*. In May, she visits the battlefront in the Crimea. Collapses from fever and exhaustion, but returns to Scutari in June. In November, the Nightingale Fund is created at a public meeting in London, raising over £44,000 from public donations.
1856	Treaty of Paris formally ends the Crimean War in March. Nightingale departs Scutari in late July, returning to Derbyshire in early August. In September, she meets

	Queen Victoria at Balmoral to discuss a Royal Commission. Nightingale moves to the Burlington Hotel, London, in November.
1857	Nightingale is busy analysing the failures of the war in various reports. In August, she suffers a second collapse and retreats into house-bound seclusion.
1858	Nightingale becomes the first female fellow elected to the London Statistical Society (later the Royal Statistical Society).
1859	Publishes *Notes on Hospitals*; in response to the Indian Rebellion (1857), Nightingale begins work on the Royal Commission into the Health of the British Army in India (completed 1863).
1860	In January, the first version of *Notes on Nursing* is published. The Nightingale Training School at St Thomas' Hospital opens in June.
1861	Death of Sidney Herbert; Nightingale suffers a further severe collapse leaving her mostly bedridden for six years.
1862	Benjamin Jowett begins conducting religious services in Nightingale's home.
1865	Nightingale moves to 35 South Street, Mayfair, her home until her death.
1872	Reform of the Nightingale Home, now in a separate building at the new St Thomas' Hospital site, and the creation of the post of Home Sister.
1874	Death of William Nightingale. Nightingale's mother moves to Lea Hurst, where Nightingale intermittently helps to nurse her until her death in 1880.
1876	The Nightingale Fund begins supporting the training of district nurses.
1893	Writes 'Sick-Nursing and Health-Nursing' at Angela Burdett-Coutts' request.
1900	Nightingale writes her final 'address' to the trainee nurses at St Thomas'.
1910	On 13 August, Nightingale dies at her home in South Street. She is buried at the Church of St Margaret in East Wellow, Hampshire.
1915	Statue of Nightingale by Arthur George Walker erected in London.

List of Figures

CHAPTER 1

Home Sweet Home?

Sweet Home: The Victorian Ideal

On Sunday 3 August 1851, Florence Nightingale went for a walk by the river Rhine. Earlier that day she had been required to greet a Prussian princess, who was visiting the infirmary at the Kaiserswerth training institution for deaconesses where Nightingale was spending the summer. In the lunchtime heat she crossed the garden by the institution's chapel, where her fellow trainees were singing.[1] The song that Nightingale heard—written by the American dramatist John Payne in 1823 and set to a melody by the English composer Henry Bishop—was one of the most popular of the nineteenth century:

> Mid pleasures and palaces, though we may roam,
> Be it ever so humble, there's no place like home,
> A charm from the sky seems to hallow us there,
> Which, seek through the world, is ne'er met with elsewhere,
> Home, sweet home!
> There's no place like home.
>
> An exile from home, splendor dazzles in vain!
> Oh, give me my lowly thatched cottage again!
> The birds singing gaily that came at my call, –
> Give me them, with the peace of mind dearer than all!
> Home, sweet home!
> There's no place like home![2]

P. Crawford et al., *Florence Nightingale at Home*,
https://doi.org/10.1007/978-3-030-46534-6_1

This song became well known in nineteenth-century society, with its sheet music selling by the thousands around the world. In the 1850s, French writers noted the phrase 'home sweet home' as a distinctively English expression.[3] The song epitomised how the late Georgian and early Victorian period constructed home as the place where the heart resided: an idealised haven of comfort, civility, and unity.

When Nightingale heard the song on that hot German Sunday in August, she wept. She did so not out of sorrow at being away from her English home and family. Rather, this marked a moment when she knew that henceforth her home lay not in her family's English country houses, but in work—as represented by Kaiserswerth and its commitment to training women in nursing and other forms of socially useful employment. 'I thought of the *home* of happy exertion, of peaceful labour which awaits us all', she wrote to her sister, Parthenope, and 'my old tears flowed'.[4] Nightingale's tears were borne of her joy and relief at finally realising that her calling was to undertake the kinds of work for which her name has endured.

Nightingale's vision of home only lying in the 'happy exertion' of hard work put her at odds, however, with the prevailing ideals of Victorian society. Since the late eighteenth century, British society had an increasingly idealised vision of the home as a place of domestic tranquillity. In part, this was a reaction to the dramatic social and economic upheavals caused by the Industrial Revolution. Domestic stability and family bonds were celebrated in part out of nostalgia, for as Beatrice Gottlieb has argued, these were thought to be 'the very values that humanity was in danger of losing in the new industrial age'.[5] The English poet John Clare highlighted a sense of alienation and disorientation associated with home and its loss in his poem 'The Flitting' (1832), which conveyed a state of being both homeless and at home:

> I've left mine own old home of homes,
> Green fields and every pleasant place;
> The summer like a stranger comes,
> I pause and hardly know her face
>
> …
>
> I sit me in my corner chair
> That seems to feel itself from home,

> I hear bird music here and there
> From hawthorn hedge and orchard come;
> I hear, but all is strange and new[6]

Home was constructed as a lost idyll—yet this ostensibly backward-looking process was in fact something new. As the nineteenth century progressed, the ideal of bourgeois domesticity exerted an ever-greater hold on British culture. Homes became places to display respectability, whether through the choice of neighbourhood, interior furnishings, or in social rituals such as afternoon tea. In many people's minds, respectability also centrally assumed an acceptance of the gendered role of women as homemakers, whose entire existence was dedicated to the fostering of a comfortable, sweet home. The concept of home became the focus of wider debates about society, morality, and economics—indissociable from arguments about women's work, or about design, style, and taste.[7] As Judith Flanders puts it, for the Victorians, '[t]he notion of home was structured in part by the importance given to privacy and retreat, and in part by the idea that conformity to social norms was an outward indication of morality'.[8]

There was an explosion of new writing about homes and households in the Victorian period, epitomised by the proliferation of magazines with names such as *The Home Companion*, *The Sunday at Home*, and *The Home Friend*. Charles Dickens's *Household Words* perhaps most clearly marked this cultural trend when, in its first issue of 1850, it announced its aspiration to 'live in the Household affections, and to be numbered among the Household thoughts, of our readers'.[9] Popular Victorian writing presented idealised home scenes of children, pets, musical instruments, reading, family meals, parties, and parlour games. The lead article of a January 1860 issue of *The Sunday at Home*, for example, was illustrated with the image of a doting wife, clinging to the shoulder of her husband, while the family cat stared into the glowing fireplace—the 'domestic altar', as architect and designer Edwin Heathcote aptly calls it (Fig. 1.1).[10] Early and mid-Victorian fiction by writers such as Dickens, Charlotte Brontë, Elizabeth Gaskell, and George Eliot further contributed to this wide-ranging culture of home, regularly featuring plots centring on themes of family, kinship, homelessness, and exile.

Art of the period further underlined the importance of domestic life and the sanctity of the family and household. Both Frederick Daniel Hardy's *Baby's Birthday* (1867, Fig. 1.2) and William Powell Frith's

Fig. 1.1 Front cover of *The Sunday at Home*, 5 January 1860

Many Happy Returns of the Day (1856, Fig. 1.3) rendered visible the notion of the sweet home. In the former, a modest but aspiring family is depicted celebrating a baby's first birthday with the lighting of a cake in the middle of the painting, surrounded by children, mother, and a cat. The fire is glowing, toys lie on the floor, writing equipment and books lie on the mantelpiece, and a fiddle hangs in the corner. At the door, the father is depicted welcoming the grandfather and grandmother, who bring a doll and basket of fruit as gifts. In Frith's painting, an altogether more luxurious tea party is underway, with the focus on an older, garlanded child surrounded by other children, parents, grandmother, and, this time, a servant bearing gifts. To the right of the image, a child kindly offers her newspaper-reading grandfather a glass while her father looks

Fig. 1.2 *Baby's Birthday* (1867), oil painting, Frederick Daniel Hardy, Wolverhampton Art Gallery, OP371

on dotingly. Both images—by no means exceptional examples—purposefully extended popular visions of families in joyful, relaxed togetherness, epitomising the Victorian ideal of the sweet home.

Needless to say, the realities of Victorian home life were frequently a long way away from this aspirational ideal. Actual homes came in all sizes depending on the wealth and social standing of their occupants, to the extent that, as David Rubenstein argued, '[t]here was no such thing as "the Victorian home"'.[11] Much urban accommodation was shared, dark, and transient—far removed from the image of tranquil domestic stability, independence, and respectability. Most people lived in uncomfortable, cramped, insanitary dwellings and were subject to high and disproportionate rents. For them, family life was, as David Kertzer and Marzio Barbagli have suggested, something of a 'perilous venture'.[12] Rapid urban expansion meant that the population of London alone rose from 1.75 million in 1831 to 4.5 million by 1901, with other cities similarly affected.[13] 'We live in muck and filthe', wrote one slum-dweller to

Fig. 1.3 *Many Happy Returns of the Day* (1856), oil painting, William Powell Frith, Mercer Art Gallery, Harrogate

The Times in 1849, describing his first-hand experience of the dire conditions that accompanied such rapid urban change.[14] How houses were used differed greatly across class lines. For the elite, the home was a place for entertaining, displaying wealth and culture, discussing politics, and developing influential networks. For many working-class people, despite industrialisation drawing increasing numbers into factories, homes were still a place of work.[15] Hundreds of thousands of domestic servants lived in their workplaces—that is, within other people's homes.

Home might not even be one place. The wealthy were often itinerant, combining town houses with multiple country residences, while many thousands lived far from what they considered home in institutions such as workhouses, prisons, lodging houses, asylums, factories, boarding schools, or military installations.[16] Others actively wanted to leave home but were constrained by social pressures or domestic duties such as caring for family members, a lack of suitable marriage partners, geographic isolation, illness, or invalidity. The most unfortunate had no shelter at all beyond that offered by the street, suffering a stark exclusion from the comforts and security of domesticity. The killer known as Jack the Ripper did not prey on prostitutes: he murdered insecurely housed women while they slept.[17]

Theories of Home

Home has long been understood as more than a mere physical structure. Historical studies have traced the home's emergence as a site of feeling, a container of personal and collective memory, and a carrier of symbolic meaning.[18] Gaston Bachelard's classic work *The Poetics of Space* (1958), for instance, offers a psychologically rich account of the way that humans use homes to rest and retire into their own familiar corners of the world.[19] The home, as the philosopher Michael Allen Fox has argued, 'stands for a place of residence, belonging, and attachment' that 'bestows familiarity, attraction, warmth of feeling, pride, a special sense of bonding'.[20]

Once the definition of home is widened beyond a private, material dwelling, it can be seen to contain larger ideologies, such as those underpinning the categories of class, gender, and race. French social theorist Pierre Bourdieu compared the house to a book in which the structure and vision of a society is inscribed.[21] Even more expansively, for the small and interconnected British elite, the 'federation of families' that Mary Abbott argued effectively governed the English nation state for centuries, the household was part of a national web of influence and power.[22] Yet the home also structured the project of nation building in a deeper sense, constructing an image of homeland that included some kinds of people and excluded others.[23] The concept of home allowed Victorian Britons to extend ideas of citizenship and nationalism into the realm of empire. Places ruled by Britain that might otherwise be seen as foreign could be endowed with a homely, civil stability drawn from the strength of the nation, allowing the many residents of the Empire, especially white residents of the dominions, to see this territory as home.[24] Many British people believed that the values they associated with home, such as order, respectability, cleanliness, and discipline, were also those that would bring about the so-called civilisation of unfamiliar places and societies.[25] Yet, on occasion, events in the Empire could jar and jolt this sense of the strong and inviolable British home. During the Indian Rebellion/Mutiny of 1857–1858, for example, *Harper's Weekly* printed an image of two Indian fighters armed with a sword and a burning torch looming over a woman and a child, a book titled 'England' nestled on their *chaise longue*.[26]

The empirical reality of the Victorian home is as varied and elusive as these symbolic meanings. It is often quite difficult for historians to piece together patchy source material about everyday living and account reliably for what went on in kitchens, sculleries, bedrooms, nurseries, drawing rooms, halls, cellars, basements, attics, lofts, morning rooms, dining rooms, living rooms, dressing rooms, sickrooms, parlours, studies, libraries, laundries, bathrooms, and lavatories. As Judith Flanders writes in *The Making of Home* (2014), 'while reconstructing the physical surroundings in which people lived is not easy, establishing how they inhabited those physical surroundings, how they used them in daily life, is even more complex and multi-layered'.[27] The difficulty is compounded by the fact that, as Flanders puts it, 'families, and homes, have always been in flux, evolving to meet the needs and circumstances of each era'.[28]

The rapid urban expansion and industrialisation of the Victorian period created just such a flux, affecting the idea and manifestation of homeliness in crucial ways. The early decades of the nineteenth century were marked by a new culture of homemaking, where filling homes with new technologies, diverse commodities, furnishings, and ornaments became increasingly fashionable, although of course, not everyone had the financial means to participate. The idea that collections of valued objects could create what Fox has called 'homeyness' and 'identity', such that material trappings transformed bricks and mortar from mere 'shells' into places of meaning, became widely accepted.[29] While some historians have understood this shift in terms of the rise of an individualist desire to reflect personal identity and taste onto material surroundings, others have placed it in a specifically religious context, arguing that Victorians overcame the exhortations of austere Protestant dissenters and Evangelicals by reframing beautiful possessions as moral objects that shaped human character and brought the divine into the home.[30] In Deborah Cohen's words: '[b]y redefining consumption as a moral act … the British middle classes sought to square material abundance with spiritual good'.[31] Homes also carried emotional associations dependent on those who breathed within its walls, retaining traces of affection or animosity between living persons, lines of continuity between the living and the dead, and even bonds between humans and their companion animals.[32]

Homes, then, are vital 'sites of history'.[33] Thinking about the home as variously a material reality, a concept, a geographical region, and a psychological impulse (the aim to create a home, feel at home, or return home) provides a sophisticated framework for thinking about aspects of the past. It offers a prism through which to consider how individual historical figures interacted both with specific physical spaces, and with the powerful ideological forces of their era.

Florence Nightingale at Home

Florence Nightingale inhabited a variety of homes throughout her lifetime. Born in a Tuscan villa, she spent much of her early childhood in Lea Hurst, a fifteen-bedroom converted farmhouse in Derbyshire, before moving in 1826 to the much grander Embley Park in Hampshire (Chapter 2). Embley thereafter became the Nightingale family's main residence, while Lea Hurst remained their summer home. The Nightingales also spent the annual spring social season living in London's Burlington Hotel. During refurbishments to Embley in 1837–1839, they travelled for eighteen months in France and Italy, taking essential elements of their home experience—such as the nanny, Frances Gale—along for the ride. In her late twenties, Nightingale undertook two extended periods of travel without her parents, to Rome in 1848–1849, and to Egypt, Greece, and Germany in 1849–1850. Then came the Crimean War, most of which Nightingale spent living out of a small room in the Barrack Hospital, Scutari, in modern-day Istanbul, Turkey. After returning from the war, Nightingale became invalided with brucellosis, suffering chronic pain and fatigue to the extent that she was frequently confined to her bed. She chose to live in London, initially returning to the familiar Burlington Hotel, to live in proximity to the politicians and influencers who could help in her work. It was only from 1865, aged forty-five, that Nightingale finally obtained a long-term home of her own (albeit paid for by her father), when she moved to 35 (renumbered as 10, late in the 1870s), South Street, Mayfair. This became her home, office, and meeting place. It was also where she later died. In the last decades of her life she rarely left this London home, except to visit her sister's marital residence at Claydon House, Buckinghamshire.

Almost all of these were upper-class residences. Nightingale's privileged background exposed her to places, individuals, and experiences beyond the reach of most contemporaries. Her large childhood homes were in beautiful settings, and her Mayfair town house situated her among elite society. Although Nightingale often challenged complacent notions of home inspired by her privileged family life, these formative experiences certainly influenced her attitudes. Central to her work ran a fascination with the condition of other people's homes and its influence upon their working lives and general health. An important feature of Nightingale's upbringing in Derbyshire, for example, was the presence of Lea Mills, a cotton factory originally constructed in 1784 by her great uncle. For Nightingale, the mill workers were a source of interest and concern, an opportunity to practise philanthropy, and to understand the health challenges facing industrial workers. It was in their cottages that Nightingale first encountered the effects of poverty and epidemic disease. Her emergent passion to care for the sick was therefore shaped by exposure to working-class health conditions that were strikingly different from her personal experiences of domestic life (Chapter 4).

Nightingale was similarly both a product of, and an exception to, social expectations for women. In the Victorian cult of domesticity, home was essentially a feminine domain. The practical, moral, and economic regulation of the household were constructed as women's duties.[34] Victorian women were also supposed to fulfil the role of carers and healers, and as Ruth Goodman points out, 'with such high rates of mortality and infection there can have been few women, as mothers, wives, sisters and daughters, who did not have to nurse someone in their family through serious illness'.[35] Nightingale enjoyed the task of nursing her own sick relatives. Yet as a young woman she found many of the other home-based social rituals of polite society confining and rejected the idea of marriage and the role of housekeeper as being her primary calling. This brought her into tension with other members of her family, particularly her mother and sister (Chapter 3). Nightingale expressly and exceptionally challenged these social structures, without entirely rejecting their influence. In her subsequent work with nursing institutions, Nightingale insisted on female-led hierarchies of authority and sought to create institutional environments in which women would feel at home in their work as an expansion rather than a substitution of the family structure (Chapter 5). This pattern of Nightingale contesting and then repurposing, but not entirely rejecting, the notions of home and domesticity borne from her social background is a recurring theme of this book.

Home had multiple meanings at different points in Nightingale's life. Home for her was a childhood sanctuary (Chapter 2), a prison constraining women's energy (Chapter 3), the nucleus of sanitary reform

(Chapter 4), a community for nurses to live and train in (Chapter 5), a discursive strategy to challenge an incompetent army command (Chapter 6), the British Empire (Chapter 7), and a spiritual refuge (Chapter 8). In these different contexts, Nightingale's relationship with home fluctuated between the sweet and the bitterly sour. In her early life, the constraints of domesticity weighed heavily upon Nightingale's ambitions to be independent and active in the world, yet she also inhabited privileged spaces of leisure, rest, education, music, and sociability. In her active career, Nightingale experienced diverse living arrangements and family-style relationships. She regularly expressed what home meant to her and habitually drew upon metaphors of homeliness to articulate her ideas of hospital or home-based health care. Her public reputation and celebrity were also bound up with motifs of homeliness and domesticity.

Nightingale's life and ideas are unusually accessible to posterity thanks to the enormous body of letters and other papers that she wrote, sent, and received during her long life—though it must always be remembered that she destroyed a large, and presumably the most revealing, portion of her papers. Study of her writings soon reveals that her notion of home involved far more than descriptions of domestic arrangements, instead encompassing many if not all the key aspects of her life and work, from her schooling and critique of social gender relations to her work with institutions, nurse training schools, India, and district nursing. At various points, home confined and gave her freedom; it was both comforting and discomforting. Home influenced Nightingale's upbringing, her approach to work, her spirituality, and her attitude to death.

This book proposes a new understanding of Nightingale by using the concept of home to bring together these various strands of her thought and life. Nightingale's attitude and relationship to home can also reveal much about developments in health and in wider cultural attitudes in the Victorian period. The ways in which Nightingale and her society understood home were fundamental to her achievements and to her public reputation as a Victorian heroine. Nightingale's case exemplifies the formidable range of cultural work that the idea of home performed in the nineteenth century.

Despite rich scholarship on Nightingale's biography and writings, including studies of her family life, there has been little attention paid to the meanings, scope, and intertwined nature of the home in her thought, life, and work.[36] In addressing this topic, this book will be the first major study of Nightingale to benefit from access to all sixteen volumes of Lynn McDonald's indispensable *Collected Works of Florence Nightingale*. It also draws on substantial research in the Nightingale family archive assembled by Nightingale's sister Parthenope and since preserved by her step-descendants, the Verneys, at Claydon House in Buckinghamshire. In analysing and contextualising this material, we draw on foundational (and still influential) work in Victorian studies that alludes to the importance of home as a concept for Nightingale herself and for the wider period. In particular, Elaine Showalter, Martha Vicinus, and Mary Poovey have interpreted home as, respectively, a structure constraining Nightingale's emancipation, an ideal transformed in the professional communities and sisterhoods of women designed following Nightingale's model, and a cultural narrative that competed with a dominant idea of military masculinity.[37] We bring these ideas into dialogue with recent work on the home, gender relations, empire, and nursing history by Alison Bashford, Deborah Cohen, Sara Delamont, Judith Flanders, Jessica Gerard, Jane Hamlett, Carol Helmstadter, Ellen Jordan, Anne Summers, Eileen Yeo, and many more. In drawing Nightingale back into the thriving, multidisciplinary arena of Victorian studies, we hope to move her beyond the often reductive controversies in which she has tended to be drawn in recent decades, such as the sensationalist focus on her 'talent for manipulation' and 'drive to dominate' in F. B. Smith's critical 1982 biography, or the comparisons to Mary Seacole, the British-Jamaican woman who ran a business and provided generous first aid to British soldiers in the Crimean War.[38] Rather than enter into these entrenched areas of debate, this book seeks to demonstrate the complexity and richness of the relationship between Nightingale and home as a concept and material reality. It offers a unique angle on an iconic woman, one that sets her thoroughly in her ideological and cultural context yet still addresses the central events of her life, legend, and enduring influence.

Each chapter will show home as a flexible, but nevertheless extremely durable, theme in Nightingale's life from cradle to grave. Home informed her vision of public health and the organisation of institutions, it influenced her sense of responsibility for the health of the citizens of Empire and, as her thinking developed, also her own concepts of spiritual wellbeing. The result was a holistic conception of health care. Humans would not be truly healthy—that is, would not flourish—unless they felt truly at home.

Notes

1. Nightingale's diary, 3 August 1851, in *The Collected Works of Florence Nightingale Volume 7: Florence Nightingale's European Travels*, ed. Lynn McDonald (Waterloo, ON: Wilfrid Laurier University Press, 2004), 527.
2. John Howard Payne, *Clari; or, the Maid of Milan* (1823; Boston: Spencer, 1856), 13.
3. Céleste de Chabrillan, *Les Voleurs d'Or* (Paris: Michel Lévy, 1857), 254. Thanks to Marc Smeets for this reference. For the significance of this phrase in terms of twentieth and twenty-first century links between health and the home see Mark Jackson, 'Home Sweet Home', in *Health and the Modern Home*, ed. Mark Jackson (London: Routledge, 2007), 1–17.
4. Nightingale to her sister, 4 August 1851, in *The Collected Works of Florence Nightingale, Volume 1: Florence Nightingale: An Introduction to Her Life and Family*, ed. Lynn McDonald (Waterloo, ON: Wilfrid Laurier University Press, 2001), 303.
5. Beatrice Gottlieb, *The Family in the Western World: From the Black Death to the Industrial Age* (Oxford: Oxford University Press, 1993), 270.
6. John Clare, 'The Flitting', in *Poems by John Clare*, ed. Arthur Symons (1832; London: Frowde, 1908), 117–118.
7. Thad Logan, *The Victorian Parlour* (Cambridge: Cambridge University Press, 2001); Hugh Maguire, 'The Victorian Theatre as a Home from Home', *Journal of Design History* 13, no. 2 (2000): 107–121.
8. Judith Flanders, *Inside the Victorian Home: A Portrait of Domestic Life in Victorian England* (New York: W. W. Norton, 2004), 24.
9. Charles Dickens, 'A Preliminary Word', *Household Words*, 30 March 1850, 1.
10. *The Sunday at Home: A Family Magazine for Sabbath Reading*, 7, no. 297 (5 January 1860): 1; Edwin Heathcote, *The Meaning of Home* (London: Lincoln, 2014), 41.
11. David Rubenstein, *Victorian Homes* (Newton Abbott: David and Charles, 1974), 12. On the material culture of Victorian homes, see Jane Hamlett, *Material Relations: Domestic Interiors and Middle-Class Families in England, 1850–1910* (Manchester: Manchester University Press, 2010); Margaret Ponsonby, *Studies from Home: English Domestic Interiors, 1750–1850* (Aldershot: Ashgate, 2007); Flanders, *Inside*; Hannah Barker, *Family and Business During the Industrial Revolution* (Oxford: Oxford University Press, 2017).
12. David I. Kertzer and Marzio Barbagli, *Family Life in the Long Nineteenth Century, 1789–1913* (New Haven, CT: Yale University Press, 2002), xxviii.

13. Rubenstein, *Victorian Homes*, 16. The wider context of working-class housing is detailed in Martin J. Daunton, *House and Home in the Victorian City* (London: Arnold, 1983); *The History of Working-Class Housing*, ed. Stanley D. Chapman (Newton Abbot: David and Charles, 1971).
14. *The Times*, 5 July 1849, 5.
15. Ruth Goodman, *How to Be a Victorian* (London: Penguin, 2014), 175.
16. On domestic living in institutional settings see Jane Hamlett, *At Home in the Institution: Material Life in Asylums, Lodging Houses and Schools in Victorian and Edwardian England* (Basingstoke: Palgrave, 2015).
17. See Hallie Rubenhold, *The Five: The Untold Lives of the Women Killed by Jack the Ripper* (London: Penguin, 2019).
18. Heathcote, *Meaning*, 22.
19. Gaston Bachelard, *The Poetics of Space*, trans. Maria Jolas (Boston: Beacon, 1994), 6.
20. Michael Allen Fox, *Home: A Very Short Introduction* (Oxford: Oxford University Press, 2016), 8.
21. Pierre Bourdieu, *The Social Structures of the Economy*, trans. Chris Turner (Cambridge: Polity, 2005).
22. Mary Abbott, *Family Ties: English Families 1540–1920* (London: Routledge, 1993), 1.
23. Alison Blunt and Robyn M. Dowling, *Home* (London: Routledge, 2006), 159.
24. On the sense of belonging to the Empire in the white dominions see John Darwin, *The Empire Project: The Rise and Fall of the British World-System 1830–1970* (Cambridge: Cambridge University Press, 2009).
25. Catherine Hall and Sonya O. Rose, eds. *At Home with Empire: Metropolitan Culture and the Imperial World* (Cambridge: Cambridge University Press, 2006), 1–31.
26. 'English Homes in India', *Harper's Weekly*, 21 November 1857.
27. Judith Flanders, *The Making of Home* (London: Atlantic, 2014), 18.
28. Ibid., 195.
29. Fox, *Home*, 77.
30. Ibid., 19, 2–3.
31. Deborah Cohen, *Household Gods: The British and Their Possessions* (New Haven: Yale University Press, 2006), 30.
32. Francis Pryor, *Home: A Time Traveller's Tales from Britain's Prehistory* (London: Penguin, 2015), xviii; Fox, *Home*, 65. From 2016 to 2019 Jane Hamlett led an AHRC-funded research project on the role of pets in British family life between 1837 and 1937. See https://pethistories.wordpress.com (accessed 20 April 2020).
33. Mary Maynes and Ann Waltner, *The Family: A World History* (Oxford: Oxford University Press, 2012), ix.

34. Kay Boardman, 'The Ideology of Domesticity: The Regulation of the Household Economy in Victorian Women's Magazines', *Victorian Periodicals Review* 33, no. 2 (2000): 150.
35. Goodman, *Victorian*, 270–271.
36. See, for example, Mark Bostridge, *Florence Nightingale: The Woman and Her Legend* (London: Penguin, 2009); Gillian Gill, *Nightingales: Florence and Her Family* (London: Hodder & Stoughton, 2004); Lynn McDonald, *Florence Nightingale: An Introduction* (Waterloo, ON: Wilfrid Laurier University Press, 2001); Lynn McDonald, *Florence Nightingale at First Hand* (London: Continuum, 2010).
37. Elaine Showalter, 'Florence Nightingale's Feminist Complaint: Women, Religion, and "Suggestions for Thought"', *Signs* 6, no. 3 (1981): 395–412; Martha Vicinus, *Independent Women: Work and Community for Single Women 1850–1920* (London: Virago, 1985); Mary Poovey, *Uneven Developments: The Ideological Work of Gender in Mid-Victorian England* (Chicago: University of Chicago Press, 1988).
38. F. B. Smith, *Florence Nightingale: Reputation and Power* (London: Croom Helm, 1982), 12–13; Seacole's story is best told in her own words: Mary Seacole, *Wonderful Adventures of Mrs Seacole in Many Lands*, ed. Sara Salih (1857; London: Penguin, 2005). For a study of the Nightingale-Seacole controversy, written by a Nightingale partisan, see Lynn McDonald, *Mary Seacole: The Making of the Myth* (Toronto, ON: Iguana, 2014).

CHAPTER 2

Childhood Homes

The Victorians thought and wrote about childhood more than those in any preceding age. Mid-nineteenth-century literature particularly teems with children, from the famous characters in Charles Dickens's *Oliver Twist* (1837) and Charlotte Brontë's *Jane Eyre* (1847) to George Eliot's *The Mill on the Floss* (1860), Charles Kingsley's *The Water Babies* (1863), and Lewis Carroll's *Alice's Adventures in Wonderland* (1865). It was a period in which conceptions of childhood were in flux. Contemporaries were conscious of the tension between idealised visions of childhood inherited from Jean-Jacques Rousseau (*Émile*, 1762) and Romantics such as William Wordsworth (*The Prelude*, 1799), and the realities of life in a world of child labour and slum housing. The Romantic view of childhood innocence was also in conflict with a stern Evangelical view—increasingly culturally dominant after 1800—of children as inherently sinful and wicked. Were children babes to be cherished, or devils in need of discipline? In Victorian fiction, childhood was seen as a period of complex emotions, of passions that shaped subsequent adulthood but perhaps also threatened it.

Florence Nightingale's childhood and family background became objects of fascination as soon as she entered the national consciousness in the 1850s. On 28 October 1854, while Nightingale and her initial group of thirty-eight nurses were still travelling to the Crimean War zone, *The Examiner* published a (not especially accurate) piece on her background, describing her as 'young, graceful, arch and popular, delighting in the

P. Crawford et al., *Florence Nightingale at Home*,
https://doi.org/10.1007/978-3-030-46534-6_2

heartfelt attractions of her home'.[1] Biographers writing during Nightingale's lifetime—following the common pattern of Victorian children's biographies, whose heroes and heroines typically showed early signs of their vocation—looked for evidence in her childhood that her nursing mission was predestined.[2] They found it in, for example, stories of her playing at nursing her dolls, or a tale recounted by a local vicar of her saving the life of an injured sheepdog when she was seventeen.[3] Nightingale's first full-scale biographer, Edward Cook, pointed out the obvious flaw with such stories: they could likely have been told of many young women of Nightingale's era. The deterministic link drawn between them and her later fame was a fallacy that obscured, rather than explained, the true reasons for her rise to prominence: 'the discovery of her true vocation ... was not the result of childish fancy, or the accomplishment of early incident; it was the fruit of long and earnest study'.[4]

Cook's biography accordingly skipped quickly over Nightingale's early childhood, but subsequent biographers have not always followed his lead in this regard. Nightingale's early life has continued to fascinate, not just because it constitutes the background to her future exploits, but for its own sake, as a useful and unusually well-documented case study of a girl growing up in an upper-class English family in the first half of the nineteenth century, with an interesting cast of characters. Ida O'Malley's 1931 biography described Nightingale's childhood in some detail, using primary source materials that have since been lost and thus are only known through her work—notably a journal and an 'autobiography' written by Nightingale in French from age eight.[5] Among the more recent books on Nightingale, the one to focus in most depth on her childhood is Gillian Gill's *Nightingales: Florence and her Family* (2004), which presents an empathetic (if occasionally speculative) attempt to imagine in detail the psychological atmosphere of the Nightingale family.[6]

In the pages that follow the focus is more on the social, ideological, and material context of Nightingale's childhood than on her personal relationships with her close family. These have been substantially covered in the biographies and can also be explored further in the first volume of the *Collected Works of Florence Nightingale*, edited by Lynn McDonald, which reproduces an extensive amount of correspondence between Florence and her family members, albeit with a heavy weighting towards Nightingale's side of each exchange.[7]

Derbyshire: Home of the Nightingale Fortune

The fortune that bankrolled the Nightingales' upper-class lifestyle in the nineteenth century and ultimately financed Florence Nightingale's career—since she refused payment for her work between 1853 and 1856 and held no official post thereafter—was accrued in central Derbyshire in the eighteenth century. The activity that built it was lead mining. Centred on ore deposits around Wirksworth, the industry had been present in the area since Roman times. In the seventeenth century, larger capitalist-style enterprises displaced traditional small-scale independent 'free' miners.[8] The industry came to dominate the local landscape, as trees were cut down for fuel and timber in the mines.[9] Wirksworth became, in Daniel Defoe's 1720s description, 'a kind of market for lead; the like not known anywhere else [except] at the custom-house keys in London', inhabited by 'a rude boorish kind of people … a bold, daring, and even desperate kind of fellow in their search into the bowels of the earth'.[10] Well before the Industrial Revolution, the region around Wirksworth, Matlock, and Cromford had the appearance and social structures of an industrial zone, and lead mining employed the majority of the non-agricultural workforce. Major investment and scientific expertise went into efforts to drain mines so that they could be exploited. Soughs to channel away water were constructed in nearby Cromford in the 1630s using plans designed by the Dutch engineer Cornelius Vermuyden, and again in 1657–1658 and 1673 using gunpowder to blast away rock.[11] New watercourses were thus created—one of which, the Cromford Sough, was used by Richard Arkwright in the 1770s to power his (and the world's) first large-scale cotton mills. Arkwright's 1771 Cromford factory—often taken as the starting point of the Industrial Revolution proper—at the time appeared as one more commercial enterprise in an area already thick with capitalist exploitation and was made possible by engineering and financing derived from the lead industry.

The local aristocracy found the mining activity aesthetically displeasing. The Duke of Devonshire strategically planted woodland to obscure industrial sites from his views at the nearby Chatsworth estate, while the Earl of Rutland escaped altogether, relocating from Haddon Hall near Matlock to Belvoir Castle in Leicestershire.[12] However, for the lesser gentry of north and west Derbyshire, the lead industry offered significant opportunities. Around 70 percent of them had a financial interest in mining after 1660, and by the eighteenth century, lead industry wealth was becoming visible in the up-to-date Georgian styles of the gentry's houses and halls, such as John Rotherham's Hall at Dronfield.[13] The peak decades for the industry (in terms of the amount of lead extracted and smelted) were in the mid-eighteenth century, following the advent of coal-fired cupola furnaces to replace water-powered smelting mills in the 1730s and 1740s. Those who invested wisely, and at the right time, could make rapid fortunes. To quote Defoe: 'the revolution in trade brought a revolution in the very nature of things … now the gentry are richer than the nobility, and the tradesmen are richer than them all'.[14]

The Nightingales were one of the families to emerge with new wealth during the life of Peter Nightingale senior (1705–1763). Little is reliably known about the family prior to the 1720s, other than that they came from humble Derbyshire origins and were nonconformist Christians. Peter Nightingale's father Thomas (1665–1734), variously described by local historians as a servant or an illiterate yeoman farmer, acquired some lead industry assets around 1700, possibly via the will of a local gentleman lead merchant, John Spateman of Wessington, thanks to his nonconformist community connections.[15] In the 1720s and 1730s, Thomas and Peter Nightingale bought property and woodland in the village of Lea, near Wirksworth and Cromford. They provided an endowment to a dissenting chapel there, as well as owning interests in at least one lead smelter.[16] Historian Stanley Chapman describes Peter Nightingale as an 'innovating entrepreneur' who experimented with the new cupola furnaces.[17] By 1754, he owned Lea Hall, the local manor house, and had renovated it, adding a five-bay Georgian frontispiece and an inscription (which can still be seen today): 'P.N. 1754'.

In the 1760s his son, also called Peter (1737–1803), enlarged the family estate by buying lands around Matlock and began to diversify his business interests. His fortune accrued during the lead boom of the 1750s and 1760s. By 1771, he had added 1381 acres to the estate, placing him among the county's leading landowners—a status reflected through his appointment as High Sheriff of Derbyshire for 1769–1770.[18] In 1776, he contributed capital to the construction of Arkwright's second cotton mill in Cromford, also renting Arkwright some land and a house.[19] In 1783, he invested £2000—a small part of his overall fortune, which stood at some £100,000 on his death in 1803—in constructing a cotton mill of his own, by the River Derwent at Lea Bridge, recruiting one of Arkwright's wheelwrights to provide technical expertise (and, like Arkwright, hiring child labour).[20] Though a small-scale operation compared to those of

Arkwright or the Strutts at nearby Belper—not to mention the steam-powered mills of Manchester that came to dominate the industry in subsequent decades—Lea Mill remained in profitable operation and in the possession of the Nightingale family until 1894. As the John Smedley Mill—the Smedleys having operated the mill as tenants of the Nightingales since 1818—it remains in operation to this day, giving it a strong claim to being the Western world's oldest extant textile factory. A hat factory built on the estate in 1793 also remained in operation into the twentieth century.[21]

Peter Nightingale had no legitimate children.[22] His will placed his estate in entail, specifying that it could only be assumed by a male descendant.[23] Thus neither his nearest relative, his sister Ann Evans, nor her daughter Mary, who had married into the Shore banking family in Sheffield, could take control of the fortune.[24] The designated inheritor was instead Mary's son—Florence Nightingale's father—William Edward Shore (1794–1874).

Homes Fit for a Gentleman

At the age of nine, William Shore had, in effect, won the lottery. The father-son continuity of the Nightingale businesses was broken, as they now passed into the hands of a boy from Sheffield with little knowledge of, or obvious interest in, how his great-uncle's lead or textile enterprises worked. During his minority, the businesses were controlled by a committee of Derbyshire trustees. When William took possession of the estate—conditional on changing his surname to 'Nightingale', which he did in 1815—he did not assume hands-on control, happy simply to collect rent from tenants like the Smedleys who knew how to operate these assets.

From William's perspective, the key possibility offered by his inheritance was not the chance to run a business empire, but the opportunity to establish himself as a gentleman of leisure and learning, at a time when Britain's landed elite looked snobbishly down on 'trade'. This social move required William to obtain a gentleman's education and manners. After boarding school, he attended the University of Cambridge, a significant choice insofar as Oxford and Cambridge at this time were 'finishing schools for gentlemen' first and places of serious learning second.[25] Nonconformist, business-minded families like the (Unitarian) Shores tended to send their sons to Scottish universities rather than to

Oxbridge if they desired them to acquire practical knowledge and qualifications, not least because nonconformists were not allowed formally to obtain Oxbridge degrees. At Cambridge, William would study classics, philosophy, literature, history, politics, and languages rather than technical or practical subjects, though he also spent some months at the University of Edinburgh in 1813–1814 taking classes in medicine.[26] He acquired a substantial amount of learning, as befitted his bookish and solitary character, but was not an outstanding student—after all, passing exams was not the main purpose of his attendance, and he carefully cultivated other gentlemanly pursuits, notably gun sports.[27] When, in 1823, he bought a second estate at Pleasley in north-east Derbyshire, one of the key attractions seems to have been the game shooting. William wrote enthusiastically to his brother-in-law that 'there is a very great show of partridges *indeed* on my new purchase at Pleasley … a very tidy Inn there & the shooting literally I believe equal to any shooting in England – 37 shots at pheasants in 1 hedge'.[28] In later years, his daughter would criticise her father's gentlemanly idleness. He 'has not enough to do; he has not enough to fill his faculties … I say to myself how happy that man would be with a factory under his superintendence with the interests of 200 or 300 men to look after', wrote Florence in 1851.[29] She may not have been wrong—but she missed the point that William's sense of self and social status was based on his ability to sustain that idleness, and distance himself from the world of factories and trade. One of his homes from home became the Athenaeum, a gentlemen's club set up in 1824 'for Literary and Scientific men and followers of the Fine Arts', with members including Lord Palmerston, Humphry Davy, Michael Faraday, Thomas Carlyle, John Stuart Mill, and John Ruskin.[30] These circles, not those of factory managers, were the ones in which William Nightingale aspired to mix.

At boarding school—a Unitarian-run institution in Epping Forest—William had befriended Octavius Smith, son of the radical Whig MP William Smith (1756–1835). William Smith was a prominent anti-slavery campaigner in the circle of William Wilberforce, and sponsor of the Unitarian Act of 1812, which extended tolerance of nonconformist preaching.[31] Based in Parndon, in Essex, Smith's family of ten children were all well-read, multilingual, and politically aware. William Nightingale's friendship with the Smith brothers led to his marriage to their sister, Frances (known as Fanny, 1788–1880) in 1818. Fanny was six years older than William and of a sociable, romantic disposition; her parents had previously refused her permission to marry a Scottish army officer on financial grounds.[32] In 1827, there was a second marriage, between William's only sibling, Mary (or Mai) Shore, and Fanny's brother Sam Smith. This was a cosy family arrangement in the sense that it ensured that the Nightingale fortune, under its male-only conditions of inheritance, would stay within the Shore-Smith families even if William and Fanny did not have any sons, as turned out to be the case. Fanny's numerous siblings included Ben Smith (1783–1860), who succeeded his father as MP for Norwich, and Joanna Smith, who married John Bonham-Carter (1788–1838), MP for Portsmouth and one of the drafters of the 1832 Reform Act. The wider family network included a number of prominent intellectual families, such as the Darwins, Wedgwoods, and Galtons. With a grandfather and two uncles in Parliament, reformist Whig politics was thus a substantial part of Florence Nightingale's family background.

In time, William Nightingale would try his hand at politics too, but it was not an immediate priority upon his marriage. William and Fanny first spent three years in Italy, on a version of the classical Grand Tour. This too was part of the traditional pattern of gentlemanly education that William was following. Protracted leisure travel around the Mediterranean became, as Grand Tour historian Jeremy Black has written, an important 'part of the ideal education and image of the social élite' in the eighteenth century, with Italy in particular seen as 'fashionable, exciting, and fairly pleasant'.[33] Although the tradition was in decline after 1800, youthful travel of this kind was still practised by aspirant liberal politicians such as Lord Palmerston and John Russell.[34] William and Fanny spent most of 1819 in Naples, Italy's most populous city at the time and seen by English tourists as a site of 'leisure, pleasure and a frisson of danger', as well as a base for exploring the nearby ancient sites of Pompeii, Herculaneum, and Paestum.[35] Their first daughter was born there and given the Greek name for the city: Parthenope. A weak child, Parthenope struggled to breastfeed after Fanny contracted a breast infection. The hiring of a wet nurse enabled her to survive—but the nurse's own child, deprived of its mother's milk, did not.[36] In 1820, the family moved to Florence, where the streets were broader, cleaner, and less crowded, and the attractions were art, theatre, and sociability (many wealthy British tourists considered Florence, of all Italian cities, to be a 'home from home').[37] They rented a villa on the city's southern outskirts, where their second daughter was born on 12 May 1820.[38] The choice of Florence as a name was, at the time, almost as unusual as that of Parthenope. Both reeked of sophisticated classicism: a world away from the sweat and dirt of Derbyshire lead miners and textile workers, who did not dream of giving their children such grandiose names.

Lea Hurst

Yet Derbyshire was nonetheless the first place that the Nightingales came to live on their return to England in 1821. The Georgian manor house of Lea Hall not being to William's taste, he decided to redevelop and enlarge (to his own designs) a seventeenth-century farmhouse on the edge of nearby Holloway village, bought by Peter Nightingale in 1771 (Fig. 2.1). By the standards of nineteenth-century aristocracy, Lea Hurst was not imposing or ostentatious, even after William's renovations in 1823–1825 had added a new north wing and a family chapel. William

Fig. 2.1 Photograph of Lea Hurst (c. 1850–1870), the Nightingale family home in Derbyshire. Frances Frith, Victoria and Albert Museum, H-X-77-A

referred to it as 'the little Hurst' and once as 'a pigmy place [*sic*]'.[39] It 'only' had fifteen bedrooms (including servants' rooms)—certainly large in comparison with the rest of the population, but not enough to host large parties of guests for extended stays, which was an important aspect of maintaining and developing upper-class social capital. An inventory of its plate, china, and glass taken in 1841 (taken because, when the upper classes moved between different houses, they would put such valuables into safekeeping with a trustworthy local) shows that Lea Hurst's kitchen held twenty-four glass goblets, thirty small saucers, eighteen dishes and teacups, and—the most numerous single item—thirty-five table knives: enough to put on a decent spread, to be sure, but not sufficient for truly lavish entertainment.[40] Nevertheless, the Nightingales had enough guests to get a lot of use out of the crockery: Fanny Nightingale's diary for October–December 1825 records the number of people staying 'in family' each night as fluctuating between thirteen and twenty-eight.[41] That number included servants; the word 'family' did not yet mean a blood-tie only, but referred to whoever happened to be sleeping under the family roof.[42]

Perched on a hillside—William wrote that 'its external park-like appearance ... seems to occupy a ridge of high land rising from the river and running into the mountains'—Lea Hurst could play the role of a country retreat, situated to make the most of views that the poet Lord Byron had compared to those of Switzerland.[43] William loved the place and once called it 'my solace and my home', regularly writing of his 'delight' at its 'beautiful colouring' and at how it would 'wonderfully light up during a day of sunshine'.[44] One of Florence's earliest letters, to her grandmother, likewise stated that 'Lea Hurst is delightful. Our gardens are full of wallflowers ... periwinkles ... monk hoods etc.'[45] Visiting relations and friends found the privacy and seclusion of 'dear little Hurstie' a charming and quaint change from grander stately homes.[46] The most eloquent description is from the novelist Elizabeth Gaskell, who in 1854 described:

> Stone-coloured walls ... trellised over with Virginian creeper as gorgeous as can be ... Down below is a garden with stone terraces and flights of steps – the planes of these terraces being perfectly gorgeous with masses of hollyhocks, dahlias, nasturtiums, geraniums, etc. Then a sloping meadow losing itself in a steep wooded descent (such tints over the wood!) to the river Derwent, the rocks on the other side of which form the first distance, and are of a red colour streaked with misty purple. Beyond this, interlacing

> hills, forming three ranges of distance; the first, deep brown with decaying heather; the next, in some purple shadow, and the last catching some pale, watery sunlight.[47]

Along with the woods and river, the flower garden, and 'fresh summery mountain backgrounds', a striking feature of Lea Hurst's setting was the area's unusual juxtaposition of beauty with industry, of genteel wealth with proletarian poverty.[48] 'How beautiful Derbyshire is', Fanny Nightingale wrote in her diary in 1846—but adding the significant rider, 'when unspoiled by improvers and manufacturers'.[49] 'Confound the railroad', wrote William during the construction of the Cromford and High Peak Railway in 1830: 'The noise is abominable and nothing but its constancy can reconcile us to it'.[50] He was, however, full of praise for the local people and especially his tenants. 'I cannot but congratulate myself on having to deal with so pleasant a set of fellows', he wrote in 1830, during a spell of torrential rain which brought hardship to local farmers. 'Even the grumblers are so merry in their lamentations over their *last penny*, that we made a laugh over our misfortunes'.[51] He seemed to enjoy the ceremony of rent days, laying on cold beef for the tenants in Lea Hurst's servants' hall and receiving them individually in the Housekeeping Room. Visiting friends would likewise (somewhat patronisingly) praise the 'nice poor people' they encountered around Lea: 'more people seem to be going in a pleasant way to Heaven in that neighbourhood than in almost any other I ever visited. There is labour and discipline, but there are so many sweet drops in the cup'.[52]

This relationship between the Nightingale family and the industrial workers in Lea informed Nightingale's later attitudes to healthcare and especially district nursing (Chapter 4). There is, however, no particular evidence that local poverty figured in her consciousness as a younger child, though she would have been aware of the interest that her mother and aunts took in the local population. Her childhood relationship to the area around Lea Hurst is instead summarised by an 1825 letter from her aunt Julia Smith:

> Dear Parthe & Flo – Thank you for your nice letters both. I like to hear everything about everybody at Lea Crich & Holloway. I love the people best, then the air, then the river, then the rocks, then the trees, then the flowers … I like to see you climb and jump about & enjoy yourselves. I hope you are neither of you afraid to ride [the pony].[53]

Lea Hurst was an idyll, but one located in an industrial area. Sequestered in privileged isolation, the inhabitants of Lea Hurst could easily romanticise the industrial lives of the nearby villagers. Any sense of the reality of working-class conditions would only have become apparent to Nightingale as she grew up and ventured into the communities. Furthermore, after 1826 the family only spent summers there, following their annual spell in London for the social season, before returning to their new permanent home for the autumn and winter—an arrangement that enhanced the sense of Lea Hurst as a delightful refuge.

Embley

Despite its charm, the Nightingales ultimately did not consider Lea Hurst a sufficient expression of their wealth and status as a family close to (if not quite part of) Britain's political and cultural elite. They therefore looked for a larger, more imposing property further south. Their first choice was Kinsham Court, near the Welsh border in Herefordshire, where Byron had lived in 1812 and 1813, but their attempts to purchase it in 1824 failed.[54] Instead, in 1825 they bought Embley Park, a 3500-acre estate in Hampshire and Wiltshire, with a Georgian mansion house at the centre.[55] The purchase price, at around £125,000, ate up the great bulk of the Nightingale fortune (William had inherited £100,000 in 1803, and the Derbyshire estate earned around £2200 per year).[56] It was enough of a stretch that guests speculated that William 'must have half of Derbyshire to enable [him] to keep up this place'.[57] A house like Embley typically required an income of £7000 a year to maintain; the Nightingale estates brought in closer to £5000. Careful management was required (though the family also had income from financial investments, which likely grew over time). Embley was nonetheless, as Parthenope later wrote, 'furnished with all that refined and cultivated taste could suggest, Greek and Italian marbles, delicate watercolour drawings, draperies of rich colour'.[58] It was filled with carved mahogany furniture, chairs of all kinds, sofas, a piano, and, later, stuffed animals and a full-size billiards table.[59] Its plate and china inventory for 1827 dwarfed that for Lea Hurst: 122 dinner plates and twenty-one 'best' tablecloths, not to mention twenty mince pie tins. Clearly, Embley enabled some serious entertaining.[60]

Whereas Lea Hurst was cosy and snug by aristocratic standards, with small rooms and low ceilings, Embley, especially after the extensive enlargement and remodelling that William implemented between 1837 and 1839, was grand and stately. Jessica Gerard's description of the typical mansion house, and what it said about its owner's social status, gives a good impression of the effect it made on visitors:

> Massive and imposing, the exterior asserted the landowner's authority, dignity, wealth and social status. The house's architecture and interior decoration, its library, fine-art collection, gardens and park affirmed his cultured taste, education and good breeding. The formal entrance, the large, opulent hall and reception rooms served as a public stage for the rituals of social performance which presented the family to greatest advantage, enhancing its prestige among social equals and exacting deference from inferiors.[61]

This description aptly describes Embley after 1839, as surviving sketches by Parthenope of the remodelled house make clear (Fig. 2.2). Some idea of its post-enlargement size may be given by the fact that the building (since further enlarged) now hosts a private school with over 400 pupils. Pamela Horn has shown how children who grew up in such

Fig. 2.2 Coloured sketch of Embley in Hampshire (1854), the second Nightingale family home. Parthenope Nightingale, June 1854, Wellcome, 571704i, licensed under CC BY 4.0

houses recalled experiencing the high-ceilinged downstairs rooms as 'vast, uncharted regions' and wandering 'among the legs of endless gilded tables in wide passages hung with tapestries and pictures' (Fig. 2.3).[62] Children were generally confined to the nursery and schoolroom on the top storey, kept away from their parents' living quarters and their guests until they were old enough to take part in social rituals.

Even if Embley was not quite on this scale when the Nightingales first moved there, it was not necessarily easy to feel at home in such surroundings. Embley would have teemed with people: constant streams of guests, but also servants, many of whom would have lived on site. There is no reliable number for the size of their staff, but an assumption of ten to fifteen full-time indoor servants is reasonable.[63] The majority of service employees were female, undertaking cooking, cleaning, laundry, and childcare (menservants earned more, and were taxed), but large

Fig. 2.3 Photograph of the living-room at Embley (c. late 1870s), once the Shore-Smiths had taken ownership. A bust of Nightingale can be seen at the back. Blanche Athena Clough, British Library, Add MS. 72840, 47

country houses were also hubs for local tradesmen, farmers, labourers, gardeners, gamekeepers, and coachmen. Dressmakers, tutors, musicians, seamstresses, and delivery boys all came and went. From the staff's perspective, Embley would have been a relatively desirable place to work, since large landowners like the Nightingales could afford to pay and treat their employees better than poorer gentry; country house servants were considered an elite within the profession. From the children's perspective, though, all this activity meant that daily life was always lived in front of outsiders, constraining intimacy and spontaneity.

A large country house and estate like Embley was, in Gerard's words, 'far more than a family's home; it was the power base of [Britain]'s political and social elite'.[64] Despite industrialisation, most wealth and power remained in the hands of large landowners until the late Victorian period. Possession of a large estate granted access to that elite, especially in the long periods of the year when Parliament did not sit and country houses were where politicians socialised. Buying an estate and 'power house' on the scale of Embley instantly turned William into a notable local figure.[65] He served as a magistrate and as High Sheriff of Hampshire in 1829, occasioning an entry in the young Nightingale's journal, recording the pleasure of seeing him ride out in state with twenty-four men to meet the judges and escort them to Winchester Cathedral.[66] It also offered a possible route into national politics. His new neighbour and hunting partner, Lord Palmerston (1784–1865) served as the (Tory) Secretary at War from 1809 to 1828 and then as Foreign Secretary in Lord Grey's Whig government after 1830.[67] To William Nightingale, the aristocratic Palmerston—who would later be Prime Minister during the Crimean War—likely appeared as the kind of late Regency gentleman-politician figure that he could model himself on, as well as a possible source of political patronage. It helped that their politics were generally aligned—like Palmerston, William supported moves to increase religious toleration and the programme of political reform that culminated in the extension of the franchise in the 1832 Reform Act.[68] The young Florence was likewise impressed with Palmerston—William noted in a letter that she 'approved very much [of] Palmerston's exposition of his foreign policy'.[69]

Historians now tend to see the 1832 Reform Act less as a progressive step towards full democracy and more as a successful defensive measure on the part of the landholding elite, not least because it helped defuse a potentially revolutionary situation in 1830–1831.[70] The Nightingales had

had their own taste of this atmosphere of unrest. Swing Rioters—agricultural labourers protesting low wages, automation, and unemployment—swept through Hampshire in November 1830, targeting landowners and clergymen. A group of around 400 descended on Embley, demanding food, drink, and money, and threatening to smash the windows if their demands were not met.[71] In the aftermath of this scare, William displayed a degree of paternalism, setting labourers to work digging on his estate—the classic paternalist work-creation response—with priority given to those with families.[72] To be on the safe side, though, in April 1831 he also joined a new Hampshire militia designed to deter future unrest, being personally sworn in by the Duke of Wellington.[73] He reassured his daughters in an 1831 letter that 'we do not expect more rioting, because the last mobs were so ignorant, & incapable of managing matters for their own interest, that prepared as people are now, other disturbances would be checked immediately'.[74] In 1840, when a group of socialists led by Robert Owen (ultimately unsuccessfully) attempted to establish a co-operative community at nearby East Tytherley, William went to investigate and sent reports on the group to a contact at Buckingham Palace.[75]

Paternalism was not in fashion among 1830s Whigs, who preferred utilitarian solutions to social problems, especially if it meant they could pay less tax. The New Poor Law of 1834, which instituted the workhouse system as a way of deterring people from claiming poor relief and reducing the burden on wealthy taxpayers, was one of their signature measures. It represented a decisive victory for Utilitarian Whigs over paternalist Tories who (for all their attachment to aristocratic government) recognised a sense of obligation to the poor, whereas Utilitarians instead espoused free-market discipline, small government, and self-reliance. William served as a Poor Law Guardian in both Derbyshire and Hampshire, indicating his sympathy with the new system. He felt that Jeremy Bentham, the great Utilitarian philosopher, had 'taught great moral truth more effectually than all the Christian divines'.[76]

In 1834, William also secured Palmerston's support for his candidacy for election to Parliament in the Hampshire constituency of Andover. He presented himself as a 'strenuous advocate for [Reform] measures' and his election literature foregrounded the Whigs' record in cutting taxes.[77] His campaign raised the prospect of an upheaval in the Nightingales' way of life: if William entered Parliament he, and perhaps the whole family, would need to spend much of the year in London, beyond their annual spell at the Burlington Hotel for the spring 'season'. His daughters were divided over the merits of this. Parthenope 'rather like[d]' the idea, but Florence lamented it: 'I had rather he would not stand at all … the alternative of living in London and breaking up our pleasant country life and our intercourse with the poor people or being separated from Papa … is what I dread'.[78] She was spared this fate: Andover returned two MPs, but her father came a narrow third, winning 100 votes out of 392 cast (a reminder of just how small the electorate remained after 1832), as the electors opted to split their representation between the Whigs and Tories rather than return William as a second Whig. In an address to his supporters, William denounced the 'notorious means'—presumably bribery—'resorted to by our opponents' as a reason for his failure, though his lack of real connection to the borough may also have been a factor. Had he succeeded, he may have become associated with one of the most notorious scandals to dog the New Poor Law: in the 1840s, inmates at Andover workhouse (established just after the 1835 election) were found to be starving, reduced to eating marrow and gristle from the rotten bones they were made to crush for fertiliser.[79]

After his defeat, William aborted his national political career—though the family remained in social contact with Palmerston—and instead devoted more of his time to the education of his teenage daughters.

Home School

Like most upper-class girls of her era, Nightingale received almost all her education at home. Both of her parents believed strongly in education. Not only had her father been to Cambridge, but her mother, Fanny, had received a good education by the standards of women of her time. Fanny's father, William Smith MP, had a library of over 2000 books and hired various tutors to instruct his daughters in music, drawing, and languages at their Essex home.[80] William and Fanny spent significant time, money, and energy on providing schools for poor children in their local areas, as part of their paternalistic duty as (in their own self-image) enlightened landowners. As early as 1808, William had contributed to a scheme to set up a school in Lea.[81] By the 1830s, he and Fanny were sponsoring schools in Lea and Crich in Derbyshire, and in Wellow, their nearest Hampshire village, as part of their philanthropic undertakings. They paid the schoolmasters' salaries and expenses and donated clothing, such as '55 pairs of cotton stockings, & 20 petticoats' to the Crich school in 1833.[82] They also sponsored the education of individual children, as shown by a note in Fanny's 1825 diary: '[p]ut to School Lea – 2 Knowles 2 Gregorys 1 Marchand 1 Woolley 1 Flint 1 Taylor; Matlock 1 Lucy Wilson; Crich 2 Hoggs'.[83] They encouraged the private tutors they hired for their own daughters to share knowledge with the local schools, as an early Nightingale letter attests: 'Mr Beber a German teacher came here [Embley] he told Susan Cromwell the mistress of our school at Wellow how to teach the children'.[84] They promoted relatively liberal methods of education in their sponsored schools, disapproving of severe corporal punishment and rote learning, and made their funding conditional on their preferred pedagogical methods being employed. In 1847, William wrote to the Duke of Devonshire, explaining that a £50 donation to the Crich school made by the Duke a decade earlier had not, in fact, been spent, since the school 'after having continued 3 or 4 years was given up, in consequence of the parents disapproving a plan of education which included play and discouraged flogging'.[85]

Local village schools were of course not where the upper classes sent their own daughters to learn. Britain's nineteenth-century education system was largely organised according to social hierarchy, with the amount of learning one received determined by one's social station; education was not intended as a means of social mobility. Upper-class girls were educated in a very different way from those of other classes, expected to reach far higher standards of literacy and refinement, though not to the point of being *too* good at anything, as Nightingale was later to complain. They were less likely to be sent to boarding school than upper-class boys, and much less likely to be taught practical or technical knowledge (though this was often lacking for upper-class children in general). As Sara Delamont has shown, upper-class girls' education was primarily conceived of as 'preparation for a flirtatious courtship'; it was thought that an education that included useful skills might drive potential husbands away.[86] At Kaiserswerth in 1851, Nightingale complained that 'my life was so wholly unpractical that I never did my own hair till I came here. I did not know the difference between rye and barley, between linen and cotton'.[87] Upper-class girls needed to attract a man and to demonstrate their ability to serve as society hostesses. As such, they were required to learn how to generate and sustain conversation on such topics as politics, literature, art, or travel, supply musical entertainment in the drawing room, and pepper their letters and conversation with witticisms and *bon mots*. As future partners of wealthy landowners, they ideally also needed the ability to manage a household of servants, maintain relationships of equal standing with other distinguished families, and reinforce relationships of deference with the local population. As mothers, they would be expected to train and discipline the appropriate moral character in their children, reinforcing the values and practices of a hierarchical society. And as members of the lesser-valued sex, they were expected to learn to express their own deference to the men they lived around. Academic learning was thus only one aspect, and not necessarily the most highly valued, of their overall formal and informal curriculum.

For the first decade of Nightingale's life, this broad social and academic education was chiefly under the control of women: ultimately her mother, but often via other figures. The Nightingale daughters had a dedicated 'nurse' or in modern terms a nanny (Frances Gale), French maids (Marguerite Agathe Selina Lanson, known as 'Agathe', and Selina Clémence Coulbeaux, known as 'Clémence'), and, from 1827, a governess, Sarah Christie. Nightingale and Parthenope also regularly stayed with aunts for weeks at a time and were thus taught by their cousins' governesses, such as Sarah Sophia Hennell, later an author and associate of George Eliot's, who was the Bonham-Carters' governess for five years.[88] Formal lessons were a relatively small part of their routine. They spent much time outside, 'grew silkworms and plants, watched caterpillars turn into butterflies, pressed flowers into albums, and enjoyed the company of creatures both tame and wild'.[89] Nightingale displayed an incipient talent for thorough classification and analysis by carefully cataloguing her flower collection.[90] She and her sister played word games, went for walks, and collected shells.

The Nightingales' engagement of Sarah Christie as governess marked the beginning of a more formal phase in their daughters' education, corresponding with the move to the statelier surroundings of Embley. As Delamont has written, the hiring of governesses could be 'a form of conspicuous consumption … a symbol of the wife/mother's emancipation from the last type of work—teaching children—and of her adoption of a totally ornamental role'.[91] Christie taught Nightingale to answer such questions as 'What was Caesar's favourite Legion? At what period was the Saxon Heptarchy completely established? What embittered the last moments of Richard I?'[92] School-style lessons took up only two to three hours of each day; emphasis was otherwise placed on well-informed conversation to encourage the Nightingale sisters 'to search for themselves'.[93]

As Gill's biography describes, however, Christie's relationship with Nightingale was not smooth. Christie came from a religious Norfolk family and espoused the Evangelical view of children as sinful beings in need of discipline. This may have clashed with the view of other influences on the young Nightingale, such as her aunt Mai Smith, who in 1827 gave her Arnau Berquin's *L'Ami des Enfans* (1782), which promoted a Romantic vision of childhood.[94] Christie appears to have decided to work chiefly on Nightingale's character and attitude, rather than on her already-evident academic ability. Nightingale likely struck Christie as too intellectually self-assured and independently minded, lacking the instinct for deference necessary to a young lady. She obliged Nightingale 'to sit still … till I had the spirit of obedience'.[95] Fanny Nightingale approved of Christie's methods and was happy to leave her daughters in her care for weeks or even months at a time.[96] She came to depend on Christie and 'lay awake two nights' upon learning of her departure to get married in 1831.[97] She did not however entrust Christie with her daughters' religious upbringing, and often herself read the Bible to her daughters before breakfast.[98]

Sympathetic observers, such as the writer and family friend Emily Taylor, could see at the time that Christie had misjudged Nightingale, and that Christie's frequent negative judgements of her character made her depressed and agitated.[99] Parthenope later wrote that Christie (who died in childbirth in 1832) 'misunderstood [Florence] completely', and Nightingale herself stated that she 'brought me up most severely. She was just and well intentioned, but she did not understand children and she used to shut me up for six weeks at a time'.[100] Gill speculates, plausibly but inconclusively, that Christie's four years with the family left Nightingale with a far-reaching sense of guilt, shaping her later tendency to excessive self-reproach and possibly repressing her sexuality.[101]

But it was not only Christie who expected Nightingale to be dutiful. As Horn has written, it was typical for the activities of country house girls to be 'strictly regulated so as to instil in them the correct moral, social and religious attitudes'.[102] Nightingale's external influences stressed the importance of such things as Bible reading, prayer, exercise, doing good to the poor, going to church, and noting the contents of sermons. The following note is from her father in 1835:

> Exercise for ten minutes every day before Breakfast. Before you dress do the exercise of the arms 20 times. In the course of the day 20 minutes exercise must be done and if not well done 10 minutes more. Run down to the gate before Breakfast by the road, or go down to the second gate upon the Poney [*sic*]. Every day you must be an hour out of doors before dinner unless you have permission to do otherwise ... Never sit down to Tea without changing, if not changed it must be taken upstairs ... Some new Poetry to be learned & two things prepared for this Evening. I always prefer varied Poetry. Practice sacred music for half an hour. Pray never omit teaching Betsy on Sunday ... If any of these things are omitted you will work them up the next day.[103]

After Christie's departure, the Nightingales considered sending Florence away to school, but instead decided to keep her at home, with her father taking over the education of his daughters. He took the task very seriously, introducing them to a wide range of challenging ideas and disciplines, including several that were not typically taught to girls, and rebuking them strongly if he felt their standards were slipping.[104] The Nightingale sisters learned Latin and Greek. They studied chemistry, geography, physics, astronomy, grammar, philosophy, French, and Italian. They discussed religion and politics. They translated medieval Italian poetry into French. They read the history of the Roman Empire and modern Europe, and wrote compositions modelled on the Oxbridge undergraduate essay for their father's judgement. They analysed and translated Plato. They studied the Bible, learned some German and even some Hebrew. At breakfast, they recited the lessons of the previous day to their father while he sat reading.[105]

This stimulating curriculum, all delivered in and with her father's library, had a transformative effect on Nightingale. In her sister's words, she 'worked patiently to the pith and marrow of every subject'.[106] In these years, during the school hours, her intellect was not limited by the shape of the society beyond the library walls but was free to advance. Notably, she was encouraged to take her learning beyond the bounds of what was necessary to become an 'accomplished woman', as Jane Austen put it in *Pride and Prejudice* (1813), and also beyond the limits of her father's education, as shown by the hiring of a mathematics tutor in 1840. William had already shown an interest in the emerging field of statistics, having been a founder member of the statistical section of the British Association for the Advancement of Science in 1833. He was on friendly terms with the statistician John Rickman (1771–1840), one of the creators and overseers of the British census. Rickman sent William details of Derbyshire's census returns in 1832, highlighting the county's relatively low death rate, and following an 1836 meeting in London praised 'your Young Ladies who receive Information with eagerness and succour'.[107] Statistical surveys, and their power to render visible correlations and patterns of behaviour that would otherwise be obscured, appealed to Nightingale's organisational mind, and later became central to her work as a sanitary reformer.[108]

Perhaps the key point about the education that Nightingale received at home with her father is that it gave her the tools and confidence to hold her own when she encountered men holding positions of power and influence in the world. In her dealings with the British government and army, she rarely if ever found herself intellectually or academically outgunned and could not be overawed or cowed into submission by some man's erudition or learning. This was an extremely unusual position for a nineteenth-century woman. It also helped that, as well as providing her with the intellectual tools, her home life had given her much experience of conversing with influential people. In this respect, a significant event during Nightingale's adolescence was Lord Palmerston's 1840 marriage to Lady Emily Cowper, which brought a new set of people into her family's social orbit. These included Richard Monckton Milnes, the most important love interest of Nightingale's life (Chapter 3). Another significant new contact was Cowper's son-in-law, Anthony Ashley-Cooper (Lord Ashley, later the 7th Earl of Shaftesbury), the Tory Evangelical social reformer and champion of the Ten Hours Acts limiting textile factory working hours. Ashley provided Nightingale with a model of how to engineer progressive legislative change through a combination of high-minded Christian motivation and indefatigable political lobbying.

These were only the most significant of a multitude of distinguished guests who came to dine at Embley. Those whose dinner acceptances were retained for posterity by Fanny Nightingale include MPs such as Edward Bunbury (also the author of a history of ancient geography), or Baron Monteagle, Chancellor of the Exchequer 1835–1839, as well as culturally significant women such as Elizabeth Eastlake, an author and art critic whose husband ran the National Gallery, or the poet Caroline Clive.[109] There was also the Prussian ambassador (and ancient historian), Baron von Bunsen, who visited Embley from 1842 and described the philanthropic work being done by women's institutions in Prussia—sparking ideas in the young Nightingale's mind (Chapter 5).[110] During her struggle to escape from conventional social expectations, described in Chapter 3, Nightingale deplored the amount of time she was forced to spend at the dinner table. But it would clearly be an exaggeration to see all this time spent at home as wasted.

Household Management

Managing Embley's bustling world of nannies, governesses, tutors, cooks, and other servants, while ensuring that the house was kept in a suitable state to feed, shelter, and impress constant streams of visitors, was clearly a complex task. Much of this work, consisting of (in modern parlance) human resources, hostelry, events management, and accounting, fell to the lady of the house, supported by a housekeeper and other senior servants. As K. D. Reynolds has shown, upper-class women's assumption of a key role in estate management could be part of a family's demonstration of its aristocratic status. Many husbands wanted their wives to understand everything about how their estates were run, so that things would still function smoothly when they were called away on political or other business.[111]

Beyond the aristocracy, the nineteenth century was famously the age of household management, perhaps associated above all with Isabella Beeton (1836–65), whose *Book of Household Management* (1861) was a famous runaway success, and who also wrote for *The Englishwoman's Domestic Magazine* (launched by her husband Samuel in 1852). For Beeton, competent management on the part of the woman of the home was essential to avoiding the 'discomfort and suffering' produced by 'a housewife's badly cooked dinners and untidy ways'.[112] Such works, aimed at the middle classes, often upheld the moral value of diligent management—as did Charles Dickens in *Bleak House*, published in 1853 after its serialisation in *Household Words* magazine. Dickens's character of Caddy Jellyby initially brings her husband to the brink of bankruptcy through bad housekeeping, but ultimately learns good household management and becomes a better person for it. By the 1850s, then, household management had become a moral, as well as a practical question. This was perhaps a little less true during Nightingale's early childhood. The anonymous but influential work *A New System of Practical Domestic Economy* (1823), sometimes attributed to Beeton's precursor Maria Rundell (1745–1828), stressed enjoyment more than virtue. It argued that 'to economise expenditure, is to unite enjoyment with prudence … true economy consists not in saving, merely, but in adapting everything to its specific use, by which much is enjoyed at small expense'.[113] The book advised on what proportion of one's income to spend on specific purposes—ten percent should go on horses and carriages, for example.

Fanny Nightingale clearly had a talent for organisation and monetary prudence. This was why, when guests queried that William must own 'half of Derbyshire' to provide the £7000 a year income necessary to Embley's upkeep, he could proudly 'answer [that] £5000 goes a long way with such management as my Missis and I make'.[114] Their daughter, who did not dispense such praise lightly, described her mother after her death as 'a most excellent manager of a household'.[115] Cook's biography states that Florence 'inherited her mother's organising capacity'.[116] The Nightingale girls were trained to account for any spending they made out of their weekly allowance (sixpence a week in 1829), such that Florence sometimes ended letters to her mother with details of money spent: 'Buns two 1 pence each; biscuits three 1 pence each' and so on.[117] By the time of her adolescence, Nightingale's organisational abilities were clear. O'Malley writes that after the renovation of Embley in 1839, she 'took an eager part in the details of furnishing and arrangements, deciding what kind of chairs would go best with the oak ceiling and pendant of the drawing-room … even helping her mother to get together a sufficient complement of junior servants'.[118] In 1842, Nightingale and her maternal cousins from the Nicholson family, who lived at the grand Surrey residence of Waverley Abbey, put on a private production of *The Merchant of Venice*; Nightingale fell naturally into the role of stage manager.[119] Later in the 1840s, to alleviate Nightingale's sense of boredom and frustration, her mother put her in charge of the still-room, pantry, and linen-rooms at Embley. Nightingale wrote in 1846 that she was 'very fond of housekeeping' as it was 'at least a practical application of our theories to something' and enjoyed making 'lists, my green lists, brown lists, red lists', as she made inventories and listed items that required replacement or repair.[120] This predilection for hands-on household organisation was to remain with Nightingale throughout her life.

Like the Nightingales' dinner parties, such practices appear in retrospect as having helped to prepare her for the life that she was to later lead: her experience of country house management influenced the way she ran healthcare institutions in the 1850s. There is less direct evidence for how Nightingale dealt with servants at Embley or Lea Hurst, but it is safe to assume that life there provided plenty of opportunity for reflection on the best methods of managing the young working-class women who made up the bulk of the staff, who came from backgrounds similar to those of many nurses that Nightingale would employ in Harley Street and elsewhere.

Nightingale clearly had a very privileged childhood. At a time of widespread child labour and horrific infant mortality, her life conditions were 'easy, comfortable, not very strenuous'.[121] She lived in plush surroundings with an attentive, loving family and interesting, intellectually stimulating visitors. She travelled, studied, played, and generally had, in O'Malley's words, about 'as rich and varied a life as was possible for little girls in pre-Victorian England'.[122] Nightingale herself acknowledged in 1851 that 'I have had experience of the best of England's life in our class'.[123] She was blessed with intellect and provided with the tools to acquire substantial learning. Her education may have been lacking in practical and technical aspects, but she was given extensive access to a wealth of art, music, literature, and culture. There were frustrations, especially as she got older, but she enjoyed large parts of her childhood.

But did she feel entirely at home? Her biographer Edward Cook argued in 1913 that what was striking about the young Nightingale was that, much as she could take pleasure in the cushy surroundings of her early life, 'her soul did not become rooted in them'.[124] Nightingale developed a feeling that her true life, the one in which she would feel fulfilled, lay elsewhere, away from the cosseted world of the country house. As her awareness grew of the world outside, with its enormous inequalities and social problems, and as she became conscious of acquiring intellectual tools that might help her find solutions, she became 'reproachful of herself for doing little'.[125] When Thomas Carlyle's *Past and Present*—an idiosyncratic and immensely influential combination of medieval history with contemporary social critique—appeared in 1843, Nightingale underlined the sections on work and copied them out in letters to relatives. The following quote stood out to her above all: 'Blessed is he who hath found his work; let him ask no other blessedness. He has a work, a life purpose: he has found it and will follow it'.[126] It was in useful, purposeful work that Nightingale felt she would truly be at home. But as the next chapter will show, finding and pursuing that work was not a straightforward task.

Notes

1. 'Who Is Mrs Nightingale?', *The Examiner*, 28 October 1854, 682–683.
2. Martha Vicinus, 'Biographies of Florence Nightingale for Girls', in *Telling Lives in Science: Essays on Scientific Biography*, ed. Michael Shortland and Richard Yeo (Cambridge: Cambridge University Press, 1996), 200.
3. Jervis Trigge Giffard, *Constance and 'Cap', the Shepherd's Dog: A Reminiscence* (London: Harrison, 1861). See Alison Booth, 'A Bestiary of Florence Nightingales: Strachey and Collective Biographies of Women', *Victorian Studies* 61, no. 1 (2018): 93–98.
4. Edward Cook, *The Life of Florence Nightingale*, vol. 1 (London: Macmillan, 1913), 15.
5. Ida O'Malley, *Florence Nightingale 1820–1856: A Study of Her Life Down to the End of the Crimean War* (London: Butterworth, 1931).
6. Gillian Gill, *Nightingales: Florence and Her Family* (London: Hodder & Stoughton, 2004).
7. *The Collected Works of Florence Nightingale, Volume 1: Florence Nightingale: An Introduction to Her Life and Family*, ed. Lynn McDonald (Waterloo, ON: Wilfrid Laurier University Press, 2001) [hereafter *CW* 1].
8. See Andy Wood, *The Politics of Social Conflict: The Peak Country, 1520–1770* (Cambridge: Cambridge University Press, 1999).
9. David Hey, *Derbyshire: A History* (Lancaster: Carnegie, 2008), 304–305. Visitors to the National Stone Centre at Wirksworth today can still see the extensive effects of mining on the area.
10. Daniel Defoe, *A Tour Through the Whole Island of Great Britain* (1724; London: Penguin, 1971). However, see also the introduction to Wood, *Politics*, which argues that Defoe misunderstood the lead industry.
11. Doreen Buxton and Christopher Charlton, *Cromford Revisited* (Nottingham: Derwent Valley Mills World Heritage Site Educational Trust, 2013), 2.
12. Wood, *Politics*, 5.
13. Hey, *Derbyshire*, 305–306, 296.
14. Daniel Defoe, *A Plan of the English Commerce: Being a Compleat Prospect of the Trade of this Nation as Well the Home Trade as the Foreign* (1728), quoted in Judith Flanders, *The Making of Home* (London: Atlantic, 2014), 43.
15. Stanley Chapman, 'Peter Nightingale, Richard Arkwright, and the Derwent Valley Cotton Mills, 1771–1818', *Derbyshire Archaeological Journal* 133 (2013): 166–188, 167. Thomas Nightingale was a trustee of John Spateman's 1708 will. DRO D37/MT/515.

16. A deed records a 1725 house purchase in Lea by Peter's father, Thomas Nightingale, for 11 pounds 5 shillings (DRO D1575/5/22). The chapel claim is made by Norman Keen, *Florence Nightingale* (Derby: J. H. Hall, 1982), 9.
17. Chapman, 'Peter Nightingale', 167.
18. Ibid.
19. Buxton and Charlton, *Cromford Revisited*, 56–57, on the Arkwright-Nightingale agreement.
20. Arkwright sued Peter Nightingale for poaching his workman, Benjamin Pearson. See Chapman, 'Peter Nightingale', 175–176; R. S. Fitton, *The Arkwrights: Spinners of Fortune* (Manchester: Manchester University Press, 1989), 224–253.
21. The Nightingale family sold the hat factory in 1903. DRO D4126/10.
22. Genealogical researcher Stuart Flint claims Peter Nightingale had an illegitimate daughter, named Mary Brown: http://www.crichparish.co.uk/webpages/stuartflint.html (accessed 8 January 2020). Peter Nightingale is often described as mad, or eccentric, in Cecil Woodham-Smith's description a 'daredevil horseman, a rider in midnight steeplechases, a layer of wagers, given to hard drinking and low company'—but no sources are provided for this. Cecil Woodham-Smith, *Florence Nightingale 1820–1920* (London: Constable, 1950), 4.
23. See Gill, *Nightingales*, 3 & n. 9 on the estate and entail.
24. Ann Evans (née Nightingale) lived in Cromford. Cromford Bridge House was purchased with her dowry. It passed to her unmarried daughter Elizabeth Evans (1762–1852) who was nursed there by Florence Nightingale, her great-niece, in her final years. It remained in the Nightingale estate until 1925. A history of the house has been written and privately printed by its current owner Pam Rivers.
25. Eric J. Evans, *The Forging of the Modern State, Early Industrial Britain, 1873–c.1870*, 4th ed. (Oxford: Routledge, 2019), 415–416.
26. Mark Bostridge, *Florence Nightingale: The Woman and Her Legend* (London: Penguin, 2008), 18.
27. Ibid., 561. William was in the fifth or sixth class out of eight or nine at Cambridge, indicating that he was something of a 'dilettante'.
28. William Nightingale to George Nicholson, 30 September 1823, Claydon N213.
29. Nightingale, note, 7 January 1851, *CW* 1, 97.
30. On the Athenaeum and other nineteenth-century gentleman's clubs, see Barbara J. Black, *A Room of His Own: A Literary-Cultural Study of Victorian Clubland* (Athens: Ohio University Press, 2012).
31. A biography exists by a direct descendant: Richard W. Davis, *Dissent in Politics 1780–1830, the Political Life of William Smith, MP* (London: Epworth, 1971).

32. See Gill, *Nightingales*, 47–51, citing documents from Claydon N18..
33. Jeremy Black, *Italy and the Grand Tour* (New Haven: Yale University Press, 2003), 1, 9.
34. See Roderick Cavaliero, *Italia Romantica: English Romantics and Italian Freedom* (London: I.B. Tauris, 2005).
35. Rosemary Sweet, *Cities and the Grand Tour: The British in Italy, c. 1690–1820* (Cambridge: Cambridge University Press, 2012), 164.
36. Gill, *Nightingales*, 57, citing Fanny Nightingale's journal, Claydon N214/1.
37. Sweet, *Cities*, 65.
38. Villa Colombaja in the parish of St Illari Podesteria dell Galliozo. The lease is held at Claydon N214/2.
39. William to Fanny Nightingale, 5 July 1830, and 7 July 1825, Claydon N213.
40. Claydon N399/3.
41. Fanny Nightingale's diary, 1825, Claydon N401/4.
42. The 1851 census defined 'members of the family' as 'the wife, children, servants, relatives, visitors and persons constantly or accidentally in the house'. Flanders, *Making of Home*, 29–30.
43. William to Fanny Nightingale, postmarked 7 February 1837, Claydon N69.
44. Bostridge, *Florence Nightingale*, 25, citing William to Fanny Nightingale, 17 January 1832, Claydon N26; William to Fanny Nightingale, 15 November 1837, Claydon N69; and 5 July 1830, Claydon N213.
45. Nightingale to her grandmother, undated [1827], Claydon N112/23.
46. See, for example, Hilary Bonham-Carter to Fanny Nightingale, 17 July 1845, Claydon N223.
47. Elizabeth Gaskell to Catherine Winkworth, 20 October 1854, cited in Cook, *Life*, vol. 1, 8.
48. Hilary Bonham-Carter to Fanny Nightingale, 17 July 1845, Claydon N223.
49. Fanny Nightingale's diary, 6 August 1846, Claydon N67/6.
50. William to Fanny Nightingale, 30 June 1830, Claydon N213.
51. William to Fanny Nightingale, 5 July 1830, Claydon N213. Original emphasis.
52. Emily Taylor to Julia Smith, 2 February 1836, Claydon N47.
53. Julia Smith to Parthenope and Florence Nightingale, undated [1825], Claydon N24.
54. Bostridge, *Florence Nightingale*, 24. Kinsham Court was later owned by the Arkwright family.
55. For the size and 1870s rental value of the estate, see the *Return of Owners of Land 1873*, a comprehensive survey of rental incomes presented to Parliament in 1875, https://catalog.hathitrust.org/Record/009024355 (accessed 22 April 2020).

56. Based on an 1820 valuation: DRO D3585/5/1.
57. William to Fanny Nightingale, 11 October 1830, Claydon N213.
58. Claydon N390; Gill, *Nightingales*, 78.
59. An 1880 inventory valued the contents at just under £3000. Claydon N111.
60. Claydon N399/4.
61. Jessica Gerard, *Country House Life: Family and Servants, 1815–1914* (Oxford: Blackwell, 1994), 5.
62. Pamela Horn, *Ladies of the Manor: Wives and Daughters in Country-House Society 1830–1918* (Stroud: Allan Sutton, 1997), 28–29.
63. Based on their income and the typical ratio of income to staff size. See Gerard, *Country House Life*, 146.
64. Gerard, *Country House Life*, 4.
65. On power houses, see Mark Girouard, *Life in the English Country House: A Social and Architectural History* (New Haven: Yale University Press, 1978), 1–12.
66. O'Malley, *Florence Nightingale*, 23.
67. Fanny Nightingale's diary shows the Nightingales and Palmerstons in social contact from at least 26 January 1827: 'N shot at Lord Palmerston's very good sport 210 head. We dined there afterwards.' Claydon N401/5.
68. He chaired the High Peak Association for Promoting Purity of Election, set up in 1831. G. J. Vernon to William Nightingale, 7 August 1831, Claydon N27.
69. Cited in Cook, *Life*, vol. 1, 6. Cook does not give a date or document reference.
70. Evans, *Forging of the Modern State*, 293–303; Toke S. Aidt and Raphaël Franck, 'Democratisation Under the Threat of Revolution: Evidence from the Great Reform Act of 1832', *Econometrica* 83, no. 2 (2015): 505–547; Carl J. Griffin, 'The Violent Captain Swing?', *Past & Present* 209, no. 1 (2010): 149–180.
71. Alex Hoss to William Nightingale, 25 November 1830, Claydon N27/2.
72. Hoss to William Nightingale, 5 December 1830, Claydon N27/3.
73. Claydon V10/167/18.
74. William to Florence and Parthenope Nightingale, 31 January 1831, Claydon N213.
75. William Nightingale to E. H. Buckley, 31 January 1840, Wellcome 9037/3, Claydon N241.
76. Cited in Cook, *Life*, vol. 1, 6.
77. Claydon V9/130.
78. Parthenope and Florence to Fanny Nightingale, undated [1834], Claydon N29/10.

79. Ian Anstruther, *The Scandal of the Andover Workhouse* (London: Bles, 1973).
80. Bostridge, *Florence Nightingale*, 14–15. Smith spent £47 in 1806 on tutoring for three daughters.
81. DRO D1575/1/21.
82. Maria Coape to Fanny Nightingale, 11 March 1833, Claydon N28.
83. Claydon N401/4.
84. Nightingale to her grandmother, undated [1820s], Claydon N112/1.
85. William Nightingale to the Duke of Devonshire, August 1847, Claydon V10/141/28.
86. Sara Delamont, 'The Contradictions in Ladies' Education', in *The Nineteenth-Century Woman: Her Cultural and Physical World*, ed. Lorna Duffin and Sara Delamont (London: Croom Helm, 1978), 134–187, 135, 141.
87. Nightingale, 'Lebenslauf', 24 July 1851, *CW* 1, 90–93.
88. Gill, *Nightingales*, 123.
89. Ibid., 92.
90. Cook, *Life*, vol. 1, 10.
91. Delamont, 'Contradictions', 136.
92. O'Malley, *Florence Nightingale*, 22.
93. Bostridge, *Florence Nightingale*, 36.
94. Ibid., 34; see also Hermia Oliver, 'The Shore Smith Family Library: Arthur Hugh Clough and Florence Nightingale', *Book Collector* 28 (1979): 521–529, here 527. The Toronto Public Library, Osborne Collection, holds a collection of Nightingale's childhood books.
95. Nightingale's journal, 15 November 1829, cited in O'Malley, *Florence Nightingale*, 24–25.
96. E.g. in autumn 1830, when she went on a health retreat in Leamington Spa. Claydon N213.
97. Mai Shore to Fanny Nightingale, undated [1840], Claydon N220.
98. Bostridge, *Florence Nightingale*, 35.
99. Emily Taylor to Fanny Nightingale, 25 February 1833, Wellcome (Claydon copy) MS9045/2.
100. Gill, *Nightingales*, 100; Nightingale, 'Lebenslauf', *CW* 1, 90.
101. Gill, *Nightingales*, 248–249.
102. Horn, *Ladies of the Manor*, 24.
103. William to Florence Nightingale, 1835, Wellcome (Claydon copy) MS9030/1.
104. William to Parthenope Nightingale, 30 April 1835, Claydon N213.
105. O'Malley, *Florence Nightingale*, 13, 38, 40; Bostridge, *Florence Nightingale*, 37–40.
106. Cited in Bostridge, *Florence Nightingale*, 38.

107. John Rickman to William Nightingale, 5 November 1832, Claydon N198; and 21 November 1836, Claydon N47.
108. On Nightingale and statistics, see M. Eileen Magnello, 'Victorian Statistical Graphics and the Iconography of Florence Nightingale's Polar Area Graph', *BSHM Bulletin: Journal of the British Society for the History of Mathematics* 27, no. 1 (2012): 13–37.
109. Claydon N106.
110. O'Malley, *Florence Nightingale*, 87.
111. K. D. Reynolds, *Aristocratic Women and Political Society in Victorian Britain* (Oxford: Clarendon, 1998), 28ff.
112. Isabella Beeton, *Mrs Beeton's Book of Household Management* (London: S. O. Beeton, 1861).
113. *A New System of Practical Domestic Economy* (London: H. Colburn, 1827), v.
114. William to Fanny Nightingale, 11 October 1830, Claydon N213.
115. Nightingale, draft letter to Julia Smith, 3 October 1871, BL Add Mss 728832a, f77.
116. Cook, *Life*, vol. 1, 7.
117. O'Malley, *Florence Nightingale*, 24; Gill, *Nightingales*, 95; Nightingale to her mother, 16 October 1828, *CW* 1, 105.
118. O'Malley, *Florence Nightingale*, 71.
119. Ibid., 80–81.
120. Nightingale to Mary Mohl, December 1846, quoted in Woodham-Smith, *Florence Nightingale*, 50.
121. Cook, *Life*, vol. 1, 7.
122. O'Malley, *Florence Nightingale*, 30.
123. Nightingale to Parthenope, 9 September 1851, *CW* 1, 306.
124. Cook, *Life*, vol. 1, 7.
125. Ibid., 13.
126. Nightingale to Julia Smith, 20 June 1843, cited in ibid., 34.

CHAPTER 3

Leaving Home

Leaving home, for an unmarried, upper-class woman in the mid-nineteenth century essentially meant exchanging one household for another. If young women from the landed classes left home for any length of time, it was generally only to go to another family property, to visit the homes of relatives or friends, or to carry out acts of philanthropic work in the homes of the poor (Chapter 4). Public outings, be they to churches and chapels, lectures and talks, or to the theatre or the opera, were always accompanied or chaperoned. Such activities presented few opportunities to engage in unplanned activities or spontaneous interactions. Foreign travel, such as the European tour undertaken by the Nightingales in 1837–1839 during extensive refurbishments to their Embley home, was only exceptionally an available option, and almost always accompanied by family members and servants from home. In the longer term, the only real way for upper-class women to leave home was to get married—which implied taking charge of a home of their own.

Leaving home outside of this limited, prescribed context was difficult for social and psychological as well as practical reasons. As we saw in Chapter 1, for the Victorians home was not simply a place but an ideology and a state of mind. For a woman to deny or refuse the benefits and comforts of stable domesticity was to risk social ostracism, being considered mentally deficient, or both. Such a worldview offered women of means—that minority of women who, like Florence Nightingale, did not need to earn a living in factories or domestic service—few socially

P. Crawford et al., *Florence Nightingale at Home*,
https://doi.org/10.1007/978-3-030-46534-6_3

acceptable functions outside of the home. Denied access to paid work, such women were instead stuck in a world of social conventions, obligations, and expectations that took up almost all their time but provided relatively little intellectual stimulation. It was from these restrictions that, as she grew older, Nightingale sought to escape. She came to feel that women's potential was being wasted, that their health and sanity were being jeopardised by desires and ambitions that they could neither fully articulate nor suppress, and which could find no outlet. As we will see below, she came to see her sister, Parthenope, as a prime example of the toxic effects that internalising conventional values could have.

Nightingale was generally fond of the places in which she grew up, especially her Derbyshire home of Lea Hurst and its surrounding countryside. Yet despite the idyllic aspects of her childhood, by the late 1840s she began to feel that she had to escape the way of life that her class background imposed on her. Her sense of alienation centred especially on her inability to carry out meaningful work. While most nineteenth-century British women worked for money, gentlewomen's social distinction consisted precisely in not needing to do so. Gentlewomen in financial need could find work as governesses or teachers without losing caste, but beyond that, the only jobs not monopolised by men were generally backbreaking and/or undesirable for women of good social repute. Nightingale came to realise that women were not able to undertake work that was truly useful, fulfilling, creative, or self-actualising. Nor were they permitted to study any subject or master any skill to an elite standard. Indeed, the notion of mastery was the order of the day in underpinning a dominant, masculine emphasis, with the domestic sphere or other social obligations blocking these ambitions. Nightingale thus came to see home as a prison, and society as a trap, preventing women from fulfilling their potential and finding useful roles in the world. Access to rewarding work became increasingly central to Nightingale as necessary for true human freedom and independence—but her upbringing had shown her that this was something generally denied to women.

If home was a prison, Nightingale began to understand the conventions of the drawing room, and the discursive and social practices of nineteenth-century society, as the bars of that prison. Dinner parties, the maintenance of correspondence, hostessing, witty conversation, the decorative arts—all were constitutive elements of a social edifice that restricted women. Such social practices worked in the interests of men, while women internalised, maintained, and reproduced their values and rhetoric. Women were encouraged to feel anxious about social status and prestige, fear what people might say if they diverged from conventional etiquette, and feel satisfaction when they acted in accordance with social norms. These pervasive values made it hard for women to break free and forge a new path. For Nightingale, the grip of social conventions and expectations brought mental torment and soul-searching. Yet she fought to free herself gradually both psychologically and physically from this incarcerated life. Her deliberations and self-questioning yielded an independent, philosophical outlook and a deep religious faith that underpinned a trailblazing career.

Nightingale was, of course, not the first or only British woman to interrogate her society's treatment of women at the time. As Kathryn Gleadle has argued, a proto-feminist movement began to emerge in the 1830s and 1840s among a network of mainly Unitarian writers and reformers.[1] The 1840s saw the appearance of proto-feminist non-fiction texts such as Marion Reid's *A Plea for Woman* (1843) and Anne Richelieu Lamb's *Can Women Regenerate Society?* (1844) as well as novels such as Charlotte Brontë's *Jane Eyre* (1847) that highlighted the generally miserable existence of governesses. Across 1840s Britain, a small but significant number of women were beginning to voice their unhappiness with social constraints and articulate the need for change, even if they had not yet coalesced into an effective collective movement—as they began to in the 1850s, when campaigners such as Harriet Taylor Mill (1807–1858), Caroline Norton (1808–1877), Barbara Bodichon (1827–1891), and Bessie Rayner Parkes (1829–1925) became prominent.

In seeking to challenge conventional expectations for her life, Nightingale can be seen as part of the emergence of this proto-feminist consciousness. Her background shared many attributes with other women in this movement, notably her strong education, political awareness shaped by family connections with leading politicians, and a degree of Unitarian influence (though Nightingale herself was Church of England). One feminist campaigner, Barbara Bodichon, was her first cousin, albeit one that the Nightingales (including Florence) considered illegitimate and therefore shunned from their company. Some of Nightingale's family members sympathised with her views—particularly her aunt Mai Smith and her mother's friend, the travel writer Selina Bracebridge. These women encouraged Nightingale's ambitions and influenced the development of her religious and philosophical outlook, helping her to outgrow most of the conventional ideologies of domesticity. Even her parents, after resisting her ambitions for many years, eventually accorded Nightingale their blessing, granting her an allowance of £500 per year to facilitate her work. Thus Nightingale, in seeking to widen the range of activities open to women of her class, and from 1849 onwards explicitly breaking free of a conventional future life of marriage, children, and country house socialising, was pushing at a door that was beginning to creak open, both in her own social set and in Britain as a whole.

Yet this incipient progress towards more liberated lives was not necessarily obvious at the time. Nightingale experienced this struggle—essentially throwing off the shackles that contained her at home—as the most formative fight of her life, taking some eight years to achieve. The starting point for this battle was her parents' refusal, in 1845, of her request to train and work as a nurse at Salisbury Hospital. The end came with her definitive departure from the parental home to take over the 'Establishment for Gentlewomen during Illness' in London's Harley Street in 1853. Creating a public life for herself cost Nightingale enormous emotional energy. She had to contend not only with the expectations of her family and wider social circle, but also with her own inevitable partial internalisation of their values. Psychologically, this was at least as much of an ordeal than anything she encountered during the Crimean War—where, despite all the external problems she faced, she was buttressed by her deep conviction that now, finally, she was doing the work she had been born to do.

This chapter will begin by examining Nightingale's strong criticism of mid-nineteenth-century landed society, and the place it afforded women, in her relatively well-known text drafted between 1850 and 1852, 'Cassandra'. It will show that Nightingale's critique, though based on her restricted and socially privileged perspective, contained some sophisticated analyses of how gendered power relations were maintained and reproduced through social and discursive pressure—analyses that in some instances anticipated the concepts of twentieth-century social theorists. The chapter will then more briefly consider Nightingale's rejection of marriage, before concluding with the crisis around her sister Parthenope that finally precipitated her definitive departure from the expected life trajectory for women of her class.

'Cassandra': Home as Prison

Nightingale wrote 'Cassandra' as one part of a much larger religious and philosophical work, which she titled *Suggestions for Thought to the Searchers After Truth Among the Artisans of England*.[2] The 'Cassandra' section (around thirty to fifty pages long depending on the edition) began life as the draft of a novel, but in its final form Nightingale dropped the dialogue and classical Mediterranean settings so that the text read more like an impassioned essay. Writing this text, and *Suggestions for Thought* in general, seems to have been a significant milestone in Nightingale's intellectual development, crystallising the emotional and intellectual distancing from her family and its values that had been taking place over the preceding years. She began 'Cassandra' following her decision, in 1849, to reject marriage to Richard Monckton Milnes, and a subsequent tour of Egypt and Greece during which she underwent much soul-searching as to her spiritual and professional vocation. Nightingale's critique of nineteenth-century society in 'Cassandra' was a narrow one, confined to the situation and experience of women of the landed classes in country homes; *Suggestions for Thought*, although theoretically addressed to the artisans of England, was also a deeply personal text. The text represents the peak of her intense frustration with life as a dutiful daughter in an upper-class home.

When Nightingale prepared the final manuscript of *Suggestions for Thought* in 1860, she sent a few copies to select contacts, but ultimately opted against publication. 'Cassandra' therefore only found a wider audience after her death, in the 1920s, when it was rediscovered by Ray Strachey and her friend (and Nightingale biographer) Ida O'Malley. Strachey published the text for the first time in 1928, with the approval of Nightingale's estate, as part of her history of the women's movement, *The Cause*.[3] Strachey's book included a chapter ('The Prison-House of Home') that discussed Nightingale's early life in the context of Mary Wollstonecraft and early nineteenth-century women's rights campaigners. Strachey saw 'Cassandra' as a key indictment of the nineteenth century's gendered social order: 'no one who reads it through can wonder any longer that women began to ask more from life than the conventions of the early years of the nineteenth century allowed them'.[4] Yet if 'Cassandra' had to wait until the twentieth century for wider recognition, it did nonetheless still have an indirect impact on nineteenth-century feminism. One of Nightingale's chosen readers for the text in 1860 was John

Stuart Mill, who was then writing what became *The Subjection of Women* (1869). As Evelyn Pugh has shown, Mill adopted several arguments that Nightingale made in 'Cassandra', and there are numerous textual overlaps between the two works.[5] Mill unsuccessfully advised Nightingale to publish her text, arguing that it was 'a testimony that ought not to be lost … an appeal of an unusually telling kind on a subject which it is very difficult to induce people to open their eyes to'.[6]

In 'Cassandra', Nightingale portrayed the country house world as a closed social system, separated from wider society. Life—for women at least—was organised according to strict norms, rules, and schedules from which no departure was permitted. Skipping the daily formal dinner, for example, or choosing to eat in one's room, was forbidden within the strangling, if unspoken, boundaries of etiquette: '[d]inner is the great sacred ceremony of this day, the great sacrament. To be absent from dinner is equivalent to being ill. Nothing else will excuse us from it. Bodily incapacity is the only apology valid'.[7] Nor was mere physical presence enough: women were expected to perform the role of vivacious conversationalist, which in turn required staying abreast of the kind of light-hearted, stimulating-yet-unserious conversation topics likely to meet with approval and fulfil the decorative role of entertainment. A young woman, wrote Nightingale, was required to 'keep herself up to the level of the world that she may furnish her quota of amusement at the breakfast table'; this performance was 'exacted from her by her family'.[8] After dinner, if her parents wanted her to converse with guests in the drawing room, she had no option but to acquiesce. When at Embley, the Nightingales 'always had company'—ten to fifteen people per night on average, as Nightingale later wrote and as her mother's diaries confirm.[9] The guests were generally politicians, writers, or military figures—'clever intellectual men, all very good society … they never talked gossip or foolishly but they took up all of our time'. Social convention in any case ensured that such conversations could never become *too* interesting: 'the essence of society is to prevent any long conversations and all *tête à têtes* … the praise of a good "*maîtresse de maison*" consists in this, that she allows no one person to be too much absorbed in, or too long about, a conversation'.[10] Conversation was required to be light, full of fleeting impressions and trifles and devoid of sustained debate—yet staying silent, or daydreaming, was not an option either. 'It is impossible to pursue the current of one's own thoughts, because one must keep oneself ever on the alert "to say something"', wrote Nightingale in the early 1850s.[11]

Fig. 3.1 Sepia wash drawing (c. 1846) of Nightingale reading beside the fireplace at Embley Park with her cousin Marianne Nicholson to her right. Parthenope Nightingale, Florence Nightingale Museum, London, 0004

In 'Cassandra', Nightingale revealed her constant struggle to carve out the space to be able to spend sustained time alone. Dinner by itself might occupy three hours per day. Social convention dictated that nothing came between women and their obligation to meet the smallest request—no 'passion' or interest should compromise the demands of 'Society'.

Women 'never have half an hour in all their lives (excepting before or after anybody is up in the house) that they can call their own, without fear of offending or of hurting someone ... we can never pursue any object for a single two hours'.[12] In a potent image of her feeling of confinement, Nightingale described being obliged to listen to a book being read aloud—a frequent, lengthy, and inescapable after-dinner drawing room experience—as being akin to 'lying on one's back, with one's hands tied and having liquid poured down one's throat'.[13] With such obligations constantly intervening, undertaking sustained study and thought was difficult for women in practical terms, and not something seen as socially valuable unless obviously linked to increasing women's social graces or marriageability.

Even in solitary moments, women found themselves beholden to the heavy demands of correspondence. For women of Nightingale's class, keeping up extended epistolary exchanges with relatives and friends was an important social obligation. Letter-writing promised an enclosed, protected space for private reflection and communication with others in comparable family situations, but, as Jane Austen's novels show, it often had to be conducted amidst the distractions of the social activities of the household.[14] Letters themselves could represent a form of social surveillance, since it was not unusual for family members beyond the author and intended recipient to read them, send them on to further readers, and discuss their contents. In 'Cassandra', Nightingale labelled correspondence as a 'bad habit': one more brick in the walls keeping women confined. Being obliged to spend time 'answering a multitude of letters which lead to nothing, from her so-called friends' was, Nightingale contended, another method by which society ensured that a young woman could not dedicate herself to independent reflection, study, or work.[15]

Nightingale could see that these various restrictions and obligations formed part of a social structure, one that served the interests of men even though it was maintained by both genders. Society, she wrote, 'uses people, *not* for what they are, not for what they are intended to be, but for what it wants them for – for its own uses'.[16] Nightingale denounced 'conventional society, which men have made for women, and women have accepted'.[17] It maddened her that, although men had constructed this ideology, women themselves had adopted it, 'have written books to support it, and have trained themselves so as to consider whatever they do as *not* of such value to the world or to others, but that they can throw it up at the first "claim of social life"'. Women had been taught, and had taught themselves, to devalue anything that was not done in the service of maintaining the structures of family, society, and church. They had closed off the option of genuine, sustained intellectual pursuits, having 'accustomed themselves to consider intellectual occupation as a merely selfish amusement'.[18] For Nightingale, this was fundamentally unnatural: a situation that only could have come about through prolonged, persistent social conditioning. Women 'are taught from infancy upwards that it is wrong, ill-tempered, and a misunderstanding of "a woman's mission" (with a great M) if they do not allow themselves *willingly* to be interrupted at all hours'.[19] They had 'no means given them' to resist such claims. Part of the maintenance of the social edifice was denying women the intellectual and emotional tools to stand up to the pressures exerted on them.

Nightingale in 'Cassandra' thus developed quite a sophisticated critique of the society around her and how it worked to maintain its power structures. Indeed, parts of her critique would strike a chord with modern sociologists. Nightingale perceived something of the ingrained attitudes, socialised norms, and impersonal forces—the 'habitus', in Pierre Bourdieu's language—by which social structures reproduce themselves.[20] She also had a sense that the discursive framework in which nineteenth-century society understood femininity was part of the problem. 'I see Mad[ame] Blanchecotte is publishing her *Impressions de Femme*', she wrote to Mary Mohl in 1868. 'What is that? Do men publish their *Impressions d'Homme*? I think it is a pity that women should always look upon themselves (and men look upon them) as a great curiosity—a peculiar strange race, like the Aztecs, or rather like Dr Howe's Idiots [the inhabitants of an insane asylum]'.[21] Such discursive treatment othered women and made them into an object of study, characterising them as deviant from the male norm. This discursive move, Nightingale saw, was an assertion of power over women, even as women themselves perpetuated it. It permitted sweeping statements to be made about them, ones that reinforced their inferior position yet should not have stood up to a moment's scrutiny. Thus, Nightingale complained in 'Cassandra', women partially perpetuated their condition by 'go[ing] about maudling to each other and teaching to their daughters that "women have no passions"'.[22] Here, again, her analysis strikingly prefigured late twentieth-century critiques of power and discourse.

It was very hard, however, to stand up to these powerful social expectations, and most women, in Nightingale's view, were not able to do so. In 'Cassandra', she characterised 'society' as an overwhelming, crushing force, one that annihilated the spirit of those who tried to resist its incessant demands. 'Society triumphs over many', she wrote. Concepts such as 'duty' and 'family' were traps. By confining women to the household domain, they restricted them to a field that was 'too narrow for the development of an immortal spirit'.[23] Young people, but young women especially, might start out with hopes, dreams, and ambitions, only to find these dwindle away under the weight and persistence of social ritual and expectation, such that their light was dulled or snuffed out:

> They wish to regenerate the world with their institutions, with their moral philosophy, with their love. Then they sink to living from breakfast till dinner, from dinner till tea, with a little worsted work, and to looking forward to nothing but bed …
>
> We see girls and boys of seventeen, before whose noble ambitions, heroic dreams, and rich endowments we bow our heads, as before *God incarnate in the flesh*. But, ere they are thirty, they are withered, paralysed, extinguished. [They are to be found] sitting in the drawing-room, saying words which may as well not be said, which could be said as well if they were not there.[24]

Though men were not immune from the harmful effects of these stultifying conventions, most of the impact fell upon women. While men too could find their dreams shattered by social reality, they had more opportunities and options. In the country house landscape, men had the freedom to retreat from society whenever they desired. They could follow intellectual pursuits seriously—indeed be admired for them—and they could reject social invitations without much fear of contestation. They had several identities open to them, including businessman, agriculturalist, politician, man of letters, romantic adventurer, huntsman, eccentric recluse, or dandyish fop. Whenever they wanted company or entertainment, the structures and rituals of country house society ensured that they could find them. As Nightingale put it, 'men are on the side of society … they say, "Why can't you employ yourself in society?" and then, "Why don't you talk in society?"'[25] Thus although day-to-day enforcement of the demands of society mostly fell to women, Nightingale, ahead of her time, recognised that ultimate responsibility lay with men; the gamut of domestic rituals and social expectations ultimately served to elevate men's goals and pursuits over those of women. Interestingly this idea, that the ideology of domesticity was a joint co-creation of the men and women of the country houses, became somewhat lost in twentieth-century scholarship that tended to ascribe responsibility for domestic life and morals more or less entirely to women, seeing nineteenth-century domesticity as something of a 'civilising restraint' imposed on men by women.[26] This interpretation changed only with the important work by Leonore Davidoff and Catherine Hall on family life in the 1980s, and John Tosh's key book on Victorian masculinity, *A Man's Place* (1999).[27] Tosh sees the ideology of domesticity as a joint creation of both genders, with 'keeping order in the home' seen as 'a critical component of masculinity'.[28] In this context, Nightingale's underscoring of the role and responsibility of men in creating the domestic ideology that surrounded her in her early adulthood is noteworthy.

As demonstrated in 'Cassandra', Nightingale's critique of the society around her, and of the structures of power, gender, and discourse that underpinned it (though Nightingale did not of course use these later critical terms), was perceptive, far-sighted, and persuasive. Her denunciation of the family gained widespread recognition within feminist scholarship of the 1980s, when Elaine Showalter followed Strachey by placing Nightingale amidst a long-standing struggle to find a hearing for the trials and sufferings of womanhood. Showalter highlighted the many obstacles that had prevented Nightingale from expressing a 'female experience as authentic and profoundly felt as that of any of our cherished heroines', including the many attempts to revise and then delay the publication of a text already adapted into a form of essay attractive to male readers.[29] However, Showalter's claim that 'Cassandra' represented a universal 'female experience' has since been questioned by the distinctly class-bound circumstances in which Nightingale was writing. While the call to escape *from* family restraints could resonate with women of all backgrounds, the allusions to what or where Cassandra might escape *to* suggest a more particular experience. The text's valorisation of a solitude normally reserved for upper-class gentlemen was understandable for those who, like Nightingale, had seen their fathers benefit from such a privilege. Writer Elizabeth Gaskell was another who fought for the liberties that her male family members enjoyed. When taking up temporary residence at Lea Hurst in 1854 (while the Nightingales were elsewhere), she celebrated sleeping in a room 'a quarter of a mile of staircase and odd intricate passage away from every one else in the house'. Enjoying relief from the duties of raising a large family, Gaskell explained that each evening she would 'lock my outer door and write'.[30]

The emancipatory ideal that afforded Gaskell and Nightingale the time and space to establish notable careers had little relevance to those for whom a writerly life was neither feasible nor desirable. In her study of subjectivity in nineteenth- and twentieth-century Britain, Regenia Gagnier discussed 'Cassandra' in relation to a middle-class tradition of feminism characterised by the valorisation of female subjectivity and meritocratic individualism. For Gagnier, this tradition prioritises the dream of 'a room of one's own' over the experience of community and co-operation that have historically been more important to working-class women's lives and movements.[31] Nightingale certainly remained cautious about wider campaigns towards female emancipation. In 1858, having left home and so relatively free from the structure that she had struggled against, she went as far as to write in a letter to Harriet Martineau that 'I am brutally indifferent to the wrongs or rights of my sex'—a comment that seems to reinforce Gagnier's identification of a damaging feminist individualism that devalues solidarity with other women seeking emancipation.[32] Similarly, despite later employing women doctors for her personal care, Nightingale complained that early female medical pioneers such as Elizabeth Blackwell 'have only tried to be "Men" & they have only succeeded in being third-rate men'.[33] Perhaps underestimating the attraction of high pay and prestige to women from less socially privileged backgrounds, Nightingale appeared to be more interested in satisfying the proven demand for female labour in nursing, rather than encouraging women to break into professional areas previously codified as male.[34] Such viewpoints do not mean that Nightingale was an anti-feminist: in her later career, she supported campaigns for women's rights (such as that mounted against the Contagious Diseases Acts), and her signature appeared at the head of an 1868 Women's Suffrage petition. Her efforts to establish nursing as a respectable and well-paying profession did much to help other women win the freedom that she herself had obtained.

The younger Nightingale's pursuit of a space apart from others in which to undertake serious work prefigures her post-Crimea working pattern, when she constructed a working space entirely under her control and beyond the reach of her family. Nonetheless, it appears somewhat at odds with her efforts in the 1850s and 1860s to forge tightly knit, committed, and disciplined communities of nurses, as explored in Chapter 5. Yet, in fact, Nightingale in 'Cassandra'—perhaps influenced by her visits to Kaiserswerth which took place around the time she was drafting this text—was already describing the desirability of creating 'institutions to show women their work & to train them how to do it', and praising monasteries as 'better adapted to the union of the life of action and that of thought than any other mode of life with which we are acquainted'.[35] Elsewhere in *Suggestions for Thought*, Nightingale proposed to repurpose the domestic structures that she criticised in 'Cassandra' in order to adapt them to professional communities. 'We would not destroy the family, but make it larger', she wrote, imagining alternative families remodelled beyond 'the ties of blood' and united by an ethic of work and the values of 'love, sympathy, protection, support'.[36] The fact that such evocative descriptions of family surface in a text otherwise famous for its rejection of domesticity reveals the complexity of the metaphorical discourses of home both consciously and subconsciously employed by Nightingale as she transitioned into her public role. Family life could constrain the subjectivity of individual women while simultaneously inspiring new models of service in the wider world. As Monica Cohen has written in an analysis of the professional and domestic values displayed in Victorian fiction, 'the domesticity conveyed in mid-century discourse [also] fostered a collectivist spirit'.[37] As Chapter 5 shows, the domestic sphere provided a plentiful discursive resource from which new public roles could be forged and justified. Nightingale rejected womanly roles in order to leave home; yet subsequently, she played upon ideals of compassion, comfort, and community in moulding new possibilities for women in the future.

Marriage Prospects

As far as conventional Victorian morality was concerned, the only way a young woman could legitimately leave her family was to get married. Marriage meant exchanging one home for another, but it did at least involve a change in status. Marriage for a woman of the landed classes implied acquiring a role in the management of a country house and estate, with significant responsibility for the hiring and firing of servants, hosting events, and organising and performing philanthropy. A woman married to a landowner had more freedom and independence than her unmarried daughters, partly conferred through the managerial responsibilities of running a large household, which also afforded her a certain amount of freedom to enter the public sphere. She was a figure of authority on her family's estates, and from the mid-Victorian period onwards, able to enter the public arena in cities without losing caste.[38] As Chapter 2 indicated, young women could be prepared for such roles by being put in charge of some aspects of their parents' household operation prior to marriage, or by being trained in the correct way to speak to servants.

Nonetheless, women typically married within their parents' narrow social set, and therefore would find themselves, after marriage, in an environment very similar to the one in which they had grown up. Marriage was a central institution by which the landed classes reproduced themselves, not just in the sense of having babies, but also in Bourdieu's sense of social and cultural reproduction: maintaining social hierarchies and behaviours, and passing on the ideas and values that preserved them to future generations. Marriage implied managing servants and holding them to upper-class expectations of deference. It meant performing the role of hostess in such a way that the husband's worldly success was advertised to his social circle. Though a society wife could take on philanthropic and, later in the century, campaigning roles, she could not stray too far from the permitted set of carefully codified social behaviours, for the sake of her husband's reputation as well as her own.

Nightingale received at least one, and probably two, offers of marriage in the 1840s. The first, less certain proposal was from her cousin Henry Nicholson in 1843–1844, not long before her first attempt to work seriously as a nurse, when she requested to enter Salisbury Hospital for a 'few months to learn the "prax"' in 1845.[39] The Nightingale daughters were close to their cousins, the Nicholsons, of Waverley Abbey; Florence in the 1840s was particularly close to Marianne Nicholson, whose husband, the engineer Douglas Galton, would later work closely with Nightingale on hospital construction projects. It is not easy to tell just how close Nightingale was to Henry Nicholson, since their correspondence was subsequently destroyed. The surviving evidence suggesting a marriage proposal is circumstantial, inferred from changes in Nightingale's relationship to the Nicholsons after 1844.[40] All that can be said for sure is that as a group, the young cousins had much amusement together within the country house confines, putting on plays, performing music, sharing books, ideas, and in-jokes about their relatives.[41] These kinds of relationships were common among the country house set, where frequent family visits of extended duration threw young people together in places otherwise isolated from the outside world. In a moral universe where relationships between non-related young people of opposite sexes were suspect, cousin marriages were common, amounting to around 10 percent of unions.[42] Such marriages were also, in Claudia Nelson's words, 'a way of safeguarding the domestic circle against change', tending to sustain the family's mores and exclude outside influences.[43] Nightingale may have interpreted any prospect of marriage to Henry Nicholson in this light; certainly, she never seriously considered marrying him. She attacked the practice of cousin marriage in 'Cassandra', contending that 'intermarriage between relations is in direct contravention of the laws of nature or the well-being of the race'.[44] As Nightingale's outlook matured, she most probably identified the Nicholsons and the 'cousinhood' with country house conventionality rather than an escape from it. Marianne in particular seems to have been somewhat scandalised by Nightingale's unconventional ambitions and in 1853 circulated a false rumour that Nightingale's parents had not given their blessing to her proposed role at the Establishment for Gentlewomen During Illness in Harley Street—an intervention seemingly aimed at scuppering Nightingale's embryonic public career.[45]

'Marriage had never tempted me', Nightingale wrote in 1851. 'I hated the idea of being tied forever to a life of society, and only such a marriage could I have'. However, 'there came a marriage for me which fulfilled all my mother's ambition, connections, everything ... I myself was tempted'.[46] This was Richard Monckton Milnes (1809–1885), a well-travelled writer and Peelite MP—he had left the Tory Party over the Corn Laws and subsequently allied with the Liberals and especially Lord Palmerston, the Nightingales' Hampshire neighbour. Milnes had met the Nightingales in 1846 and remained in close contact until Nightingale refused him in 1849. (They remained on good terms thereafter: Milnes in the 1850s was a strong supporter of Nightingale in Parliament and a trustee of the Nightingale Fund.) It is not hard to see the basis of the attraction. Nightingale and Milnes shared intellectual interests, for example attending together (with Nightingale's father) a meeting of the British Association for the Advancement of Science at Oxford in 1847.[47] Nightingale was impressed by Milnes' concern for Irish famine victims and interest in the French writer George Sand.[48] Both were well travelled in Europe and around the Mediterranean. They were equally passionate about European politics, then going through a turbulent period around the revolutions of 1848. They also both thought deeply about the future direction of spirituality in Britain, Milnes having been a contributor to the Tractarian debate over the future of the Church of England. Both took an interest in Catholic doctrine, Nightingale particularly so after a visit to Rome in 1847–1848 during which she had an intense religious experience in the Sistine Chapel, and spent ten days in a convent. (This trip also allowed her to meet several people who were to become important in her subsequent life, notably Sidney and Elizabeth Herbert and [later Cardinal] Henry Manning.)

The two appeared, in other words, very well matched. Milnes was well able to assist and encourage Nightingale's intellectual development and provide an abiding source of stimulating conversation. She knew, as she wrote, that her 'intellectual nature' and 'passional nature' that 'requires satisfaction … would find it in him'. Yet Nightingale could not escape the doubt in her mind that marriage, even to Milnes, might end up meaning simply 'a continuation and exaggeration of my present life, without hope of another'.[49] Milnes might believe in the advancement of women, but in practice his political role, and influential position within intellectual and literary circles, would necessitate partaking in a good deal of conventional sociability and attending a lot of dinner parties. He would need a wife willing to play the role of society hostess on a regular basis and with good grace. Where would that leave Nightingale and her own ambitions? It would, she felt, 'be intolerable to me'. She therefore chose instead to become one of the 'few' that she described in 'Cassandra', who 'sacrifice marriage, because they must sacrifice all other life if they accept that'.[50]

Nightingale was in little doubt as to what she was giving up. She was twenty-nine. A better-matched suitor than Monckton Milnes coming along was hard to imagine, and even if one had, she would almost certainly have faced the same dilemma. In refusing him in 1849, Nightingale had effectively decided her course in life as an unmarried woman. The problem thus became how she would gain her family's permission to leave home in another way. Help came fortuitously, from family friend and travel writer, Selina Bracebridge, who Nightingale had already accompanied (along with her husband Charles) on the 1847–1848 trip to Rome. In the aftermath of her rejection of Milnes, Nightingale went with the Bracebridges on a ten-month tour of Egypt, Greece, and Central Europe. This provided a period of separation from immediate family and introspection for Nightingale, during which time she wrote quite intensively—her letters from Egypt were privately published by her family in 1854 and have become an important reference point in understanding mid-century British orientalist attitudes.[51] Those attitudes were of course inflected by Nightingale's profound questioning of domestic life and women's role in Britain at this time. Afforded special access to observe how women were confined in another social context, Nightingale's letters struggle against acknowledging the clear connections to the situation she was facing at home. Egyptian women were similarly caged by family life, but Nightingale's imperialist outlook prevented her from drawing a direct comparison between her trials and those of her Middle Eastern counterparts.[52]

On this trip, Nightingale referred repeatedly to being tormented by daydreams, and experienced several further religious visions, the themes of which can often be clearly related to her thwarted desire to work:

> And the last spirit answered, Life is no valley of tears, that thou shouldest go through it, as through a desert which thou must traverse, bearing and waiting whilst in the world there is evil; life is none of these things. What is life then? I cried. And all the colours seemed to form themselves again into one white ray, and their voices to become one like the voices of the wind, and to say life is a fight, a hard wrestling, a struggle with the principle of evil, hand to hand, foot to foot – not *only* in thyself, nor *only* in the world, but in thyself … The rainbow had vanished, the sun was setting, and I kneeled before it and said: 'Behold the handmaid of the Lord', give me tomorrow my work to do.[53]

If the last allusion identifies Nightingale with the Virgin Mary ('the handmaid of the Lord'), the reference to traversing the desert calls to mind Jesus himself. On her thirtieth birthday, spent in Athens, Nightingale wrote to her mother that her youth had been 'misspent' and noted that thirty was the age 'when our Saviour began his more active life'.[54] On her leisurely return in 1850 via Vienna, Prague, Dresden, Berlin, and Hamburg, she visited art museums but also various charitable and medical institutions, meeting their founders and observing their work.[55] Most famously, she spent two weeks at the Deaconesses Institute at Kaiserswerth, in the Rhineland, and wrote up a detailed account of life there, as Chapter 5 will explore.

The fact that Nightingale's parents countenanced—and financed—such a lengthy and formative voyage, following her rejection of an apparently excellent suitor in Monckton Milnes, is a reminder that Nightingale's position in such a wealthy and cultured family brought her opportunities as well as restrictions. The trip represented an extended chance to pursue learning and introspection at leisure and gave her a rare period of time where she was more or less unhindered by social obligations. By the time she re-joined her family in Derbyshire in late summer 1850, Nightingale had a much clearer idea of the life she wanted to lead and a more practical sense of what this work would involve. Her family, however, expected her now to resume her former life; to take up her place in the drawing room as if she had never been away. 'Cassandra' was written in the atmosphere of tension and frustration that resulted. In the end, it was her sister Parthenope who, inadvertently and unwillingly, provided Nightingale with the justification she needed to make the final break.

Parthenope: The Sister Who Stayed at Home

Nightingale's elder sister Parthenope (1819–1890), later known as Lady F. P. Verney, was in many ways a more typical female representative of the late Regency/early Victorian landed classes than Florence. As noted above, the alienated women who made up the emerging 'slow-burn feminist wave' of the 1840s and 1850s were a small minority.[56] Many upper-class daughters grew up conforming to their parents' hopes and plans—evidence for the general effectiveness of their education in reproducing social values and structures. In Florence's eyes, Parthenope, or 'Parthe', was the archetype of a young woman who had internalised her society's values. 'She is in unison with her age, her position, her country', Nightingale wrote in 1851. 'She has never had a difficulty, except with me. She is a child playing in God's garden and delighting in the happiness of all His works, knowing nothing of human life but the English drawing room, nothing of struggle in her own unselfish nature'.[57] Parthenope, in her sister's account, appears dainty, dreamy, unserious, unperturbed by the restraints placed on her by social convention since 'she has not the smallest ambition'. She was good at drawing, languages, painting, poetry—the decorative arts—and hopeless at maths and logic. She was, thought Florence, 'the flower, the whole flower and nothing but the flower'.[58]

Nightingale's view of her sister as conventional, frivolous, and uncritical has influenced subsequent representations of her, both in biographies and in more popular versions—such as the 1951 film *The Lady with a Lamp*, in which Parthenope appears as empty-headed, inane, and trivial, discussing flower arrangements while Florence saves the lives of British troops. Such representations are somewhat unfair. Parthenope, as her letters make abundantly clear, was intelligent, erudite, perceptive, and witty.[59] She was a skilled sketch artist: many of the famous images of the young Florence are by her, such as the Wellcome Collection's image of Nightingale with the owlet, Athena, that she brought back from Greece in 1850 (Fig. 3.2). As Lady Verney, Parthenope later became a published novelist of reasonable repute, writing several romance novels employing regional vernacular dialogue in a manner reminiscent of Walter Scott. The heroine of her first novel, titled *Stone Edge* (1868) and set in the

Fig. 3.2 Drawing of Florence Nightingale stood next to a pedestal with Athena, her pet owl. Parthenope Nightingale, 1855, Wellcome, 7446i, reproduced in Sarah Tooley, *Life of Florence Nightingale* (London: Bousfield, 1904), 208

Derbyshire countryside, was named 'Cassandra', though the autobiographical resonances are otherwise limited. Parthenope also published a volume of essays on European peasant proprietors and a history of the Verney family during the English Civil War.[60] The latter work, completed after her death by her step-daughter Margaret Verney, remains a significant source for Civil War historians.[61] Parthenope, in other words, was certainly no airhead. It was just that, unlike her sister, she aspired to be successful in ways that did not challenge contemporary expectations.

Landed families had disproportionately high numbers of individuals who never married. This was true of 28% of women of Nightingale's generation (born 1800–1824), rising to 40% of girls born later in the century, compared to 12% of the female population as a whole.[62] It was thus not uncommon for one daughter, often the eldest, to be earmarked to stay at home in order to be a comfort to her parents. Parthenope appeared to be the clear candidate for this role in the Nightingale family (one which Anne D. Wallace has sought to recuperate as 'primary and generative' rather than, as is more commonly supposed, 'contingent and dependent').[63] Parthenope was often ill—including a serious fever in 1836 that caused her parents to fear for her life—and thought of as frail and sensitive, arousing protective emotions in her close relatives.[64] She was shorter in stature than Florence and considered both less pretty and less intelligent. After meeting the Nightingales in 1846, Richard Monckton Milnes reported to his sister that 'my feeling as to the two is very much that I am conscious of condescending to the one and of looking up to the other, which is true physically as well as morally' (Fig. 3.3).[65] Thus, the division of labour between the two daughters was apparently clear: Parthenope would stay at home, helping her parents with their social calendar and eventually looking after them in their old age; Florence would be the one to make a suitable marriage and carry on the family line.

Parthenope's presence played an important psychological role for Florence, giving her something to both measure herself against yet also distance herself from. As she became more conscious of her desire to seek an unconventional life, it became helpful for Nightingale to think of her sister as the epitome of conventionality, emblematic of the social edifice holding her back. There is evidence that, from an early age, Nightingale was using Parthenope as a kind of anti-model, defining herself and her personality in opposition to her sister. In the early 1830s, the Nightingale girls spent some time with Emily Taylor, a friend of their aunt Julia Smith. Taylor ran a school in Norfolk and subsequently maintained

Fig. 3.3 Watercolour painting of the Nightingale sisters (c. 1836). Florence is seated to the left and concentrating on her embroidery while Parthenope stands over her with a book, in an arrangement that may have been deliberately calculated to undermine the typical view of the sisters as expressed by observers such as Monckton Milnes. William White, © Copyright National Portrait Gallery

a correspondence with both girls. She found Florence insistent on the idea that '"Parthe and I are so different, that we require quite different treatment"', writing to Fanny Nightingale, their mother, that

> I am not very easy in corresponding with [Florence] since I find so much disposition in her to detach herself from her sister in everything … I do not find that I can conscientiously indulge her in this … speculation, in words, on the differences between her and Parthe [is] very bad indeed for her.[66]

This theme of seeking differentiation from her sister as a method of constructing her own identity recurs in Nightingale's life and writing. In a note from the 1840s, she wrote that Parthenope 'is like the Bird of Paradise, who floats over this world without touching it, or sullying its bright feathers with it, rather than the nightingale, which makes its nest in it and sings'.[67] Thus she, Florence, was the one true Nightingale, by virtue of her desire to engage meaningfully with the world's problems, which her sister apparently did not share. In 1861, by which time she had achieved a considerable degree of public impact, she asked her father rhetorically, 'Where should I have been now in any part of my life's work, had I followed any part of her [Parthenope's] life's advice?'[68]

It helped Nightingale's own sense of identity to think of Parthenope as part of the forces holding her back. Parthenope certainly did try to prevent Florence from leaving; yet this was perhaps more through fear than a desire to enforce conventional values. Her behaviour in the early 1850s makes most sense if it is assumed that, from her point of view, life as an (apparently terminally) unmarried daughter in a country house was tolerable so long as she had her sister to share it with. She could not, however, cope with the idea of living it on her own. Florence's desire to leave was therefore something to stifle at all costs. In 1851, Nightingale returned to Kaiserswerth for further training and experience, writing back to her family that 'I find the deepest interest in everything here and am so well, body and mind. This is life – now I know what it is to live and to love life and really I should be sorry now to leave life'.[69] Nightingale later wrote to Henry Manning that after her return from this trip, Parthenope had refused to discuss Kaiserswerth or even hear the name mentioned in her presence.[70] She could not stand the notion that she might be losing her sister to Kaiserswerth, or anywhere else.

Matters came to a head in autumn 1852, when Nightingale undertook another trip to develop her experience of nursing and hospitals—this time to St Vincent's Hospital in Dublin, run by the Catholic Religious Sisters of Charity. The prominent Catholic convert Henry Manning, with whom Nightingale had been corresponding about her religious views over the summer (Chapter 8), offered to help with the arrangements. To Parthenope, the Ireland trip must have seemed a portentous development. What if Florence became a Sister of Charity and never came home?

Nightingale left for Ireland on 19 August 1852.[71] On 7 September, she wrote to Manning from Belfast. She had not yet been accepted as a temporary worker at St Vincent's, but intended to return to Dublin to try again.[72] However, the following day, she was instead summoned away, to Scotland, by a family letter informing her that her departure to Ireland had occasioned a collapse in Parthenope. Parthenope appears to have had some sort of nervous breakdown. Her parents were concerned enough to send her for treatment with Sir James Clark, Queen Victoria's personal doctor, which meant travelling to Aberdeenshire, as the royals were at Balmoral. Nightingale answered the summons and went to meet her there, finding her 'delirious, though knowing me'.[73]

It was immediately clear in Nightingale's mind what had happened. Her sister had suffered 'the downward course of the finest intellect and the sweetest temper, through irritability, nervousness and weakness, to final derangement, and all brought on by the conventional life of the present phase of civilisation, which fritters away all that is spiritual in women'.[74] This echoed her indictment in 'Cassandra' of 'conventional idleness', which brought about an 'accumulation of nervous energy' in women that 'makes them feel every night, when they go to bed, as if they were going mad'.[75] In another 1850 text, she had described women 'who are suffering from ill health, merely from having nothing particular to do'.[76] Nightingale now made an ally of Sir James Clark, 'who', as she wrote to Manning, 'has been to me like a father', reporting his medical opinion that 'imbecility or permanent aberration is the inevitable consequence, unless my sister is removed from home and placed under a firm and wise hand'.[77] Furthermore, 'the medical men are decidedly of opinion that my presence at home aggravates the disease'; Clark had apparently diagnosed Parthenope with having a 'monomania'—then a relatively new and fashionable French-import diagnosis translating roughly as an unhealthy obsession, or a form of insanity restricted to a particular topic—about Florence.[78]

It is not clear from Nightingale's letters whether she herself, in conversation with Clark and his colleagues, might have had any influence over this diagnosis. The judgement emanated from a mid-Victorian medical profession that, as Showalter has shown, treated mental illness according to prevalent ideas about gender and domesticity.[79] Sally Shuttleworth has argued that mid-Victorian fears around insanity increased in part because new medical theories emphasised that everybody carried within them the possibility of becoming insane (and that the insane could be cured), rather than lunatics being a distinct outcast group requiring incarceration. If there were no obvious outward manifestations of madness, then in practice the test of sanity became self-control and obedience to designated social roles. Diagnoses like monomania, or its English counterpart, 'moral insanity', were, according to Shuttleworth, 'symptomatic of the tightening networks of social control in the Victorian era as the inner self becomes the target of ideological surveillance'.[80]

Given that it was Florence, not Parthenope, who was challenging the designated roles, it ought to have been she who was most at risk of being labelled insane. Indeed, Nightingale wrote in 1851 of her mother's 'disappointment in me … as if I was becoming insane when she has organised the nicest society in England for us, and I cannot take it as she wishes'.[81] But Nightingale was otherwise always clear that she was the sane one, and that it was the conventional society around her that was mad. In 1850–1851, after returning from her extended travels, Nightingale had briefly worked at an evening school for women factory workers, giving it up at her mother's request after (in Nightingale's later account) 'my sister went into hysterics'.[82] Nightingale's subsequent verdict was that her acquiescence to this pressure had been 'an act of insanity … had there been any sane person in the house, he should have sent [lunatic asylum superintendent Dr John] Conolly to me'. In this context, it is perhaps not surprising that Nightingale should have experienced Parthenope's monomania diagnosis as both a liberation and a vindication. What better proof of her contention that conventional family life was toxic for young women could there be than the collapse of her sister, whose life was so perfectly in tune with conventional values? 'You asked me whether I had anticipated this', she wrote to Manning, '– Oh! For such *long, long* weary years have I been expecting it that it is almost a relief it has come at last'.[83] In a comment that suggests that Nightingale was seeking to direct the emerging narrative around her sister's illness in a direction that would be useful to her, she added 'in your direction of young ladies, it may really be of some use to know what certain modes of life will lead to'.[84]

Even better, from Nightingale's point of view, was that she now had a prestigious doctor clearly stating that her presence at home was bad for her sister's health: Parthenope needed to be away from the family, and especially Florence, in order to recover. Parthenope's frailty and sensitivity, in other words, could no longer be used as a pretext to keep Florence at home. For everyone's benefit, she had to get away. She did offer to take Parthenope with her, to take full charge of her 'away from home, at any place the medical men may name', but probably did not seriously expect this offer to be accepted.[85] The irony that Nightingale's final break with home could occur only once legitimated by the advice of a male professional indicates the complex dynamics in which she was constantly embroiled. On the one hand, Nightingale was prepared to radically shake up social expectations; on the other hand, she would compliantly conform to elements of the existing system when it suited her needs and desires.

From this point on, Nightingale's final departure from the family home was just a matter of time and details, a question of choosing the right opportunities. When her Aunt Evans died in late 1852, Parthenope promoted the idea of turning Evans's old house, at Cromford Bridge in Derbyshire (two miles from Lea Hurst), into a nursing home under Florence's control. This was, on the face of it, a good compromise: an opportunity for Florence to fulfil her professional ambitions while staying close to the family. But this was not what Nightingale had in mind, and her friend Selina Bracebridge was quickly deployed to kill that idea. 'I know your heart is very sore', Bracebridge wrote to Parthenope in early 1853. 'I think of you often with the truest sympathy and admire the courage with which you think of and promote plans, which must be so trying for you … [but] her path, if she went there at once would be a very difficult one'.[86] Nightingale wanted more training and experience, to be in environments where she could develop her skills. Running a small institution by herself in a provincial town would not allow her to do this. There would be practical difficulties in starting from scratch, and adopting a family home for her work would hardly provide her with psychological distance from the world she was leaving behind. In February 1853, she left for Paris, to train with the Sisters of Charity. Although the death of her grandmother intervened to briefly bring her back, she would never truly return home again, at least not in the sense that her family understood the term. From then on, Nightingale's life would be defined by her work.

Home Both Constrains and Enables

Nightingale's relationship with her family and home was riddled with paradox and ambivalence. Home and family constrained her, in all the ways that she laid bare in 'Cassandra', but they also enabled her. Her family's wealth, education, and contacts gave her opportunities open to few others. Her education, as Chapter 2 showed, provided her with an unusually sharp and varied intellectual advantage. Her family's wealth meant that she was surrounded by the gendered values of the upper classes, but it also offered a sense of security that facilitated the tolerance of unconventional thinking. It was only in this environment of privilege that she could refuse Richard Monckton Milnes, since she was not in danger of ending up impoverished if she remained unmarried. She had the opportunity instead to take a year out, visit the archaeological treasures of the Mediterranean, and think deeply about her philosophy, theology, and identity. Nightingale felt that her home was a prison. Yet it was a prison that nevertheless provided her with the means and tools of escape.

Nightingale loved her close family members and remained in contact with them for as long as they were alive. Several members of her extended family, notably her aunt Mai Smith and cousin Hilary Bonham-Carter, worked with her closely in her later career. Leaving home did not mean cutting all connections, but it did mean that she increasingly sought to establish contact with her family at arm's length and on her own terms, so as not to feel smothered by anxieties and obligations. Nor did she dislike the houses and communities in which her family lived—indeed, she nurtured and treasured her memories of her home landscapes right up until her death. When Nightingale left home, she was not leaving places or people so much as a way of life and a state of mind: one in which women could not study seriously or engage in intellectually challenging work, but must always be prioritising appearances, social etiquette, status, and reputation. Yet even here, her possession of these social graces, learnt at home, came in useful when her later career obliged her to work with men (primarily) and women of wealth and power, such as Lord Palmerston, Sidney Herbert, or Lady Canning. Nightingale had to leave home in order to work, yet home sustained that work, through the financial, administrative, and emotional support that she received, or even in the more literal sense of the food parcels her mother sometimes sent from Embley.[87] Crucially, home provided a set of values and practices from which a public life could be constructed and legitimised—both for herself and for many more women besides.

Home, in this chapter's sense of a place in which upper-class young women's freedom of action was severely constrained by social convention, was represented by Nightingale as an unhealthy place. It was unhealthy not so much in the sense of being dirty or unsanitary, but in that it was harmful to well-being, destructive of intellectual independence, crushing of personal identity, and conducive to depression, hysteria, and lassitude. She saw the comforting stories English people told themselves about the benefits of female-curated domesticity as a myth: 'Oh poor John Bull, don't think, as you are (and will be) told every day, that "nowhere are there such homes and such mothers as in England"'.[88] The nervous breakdown of her sister Parthenope, who was guilty of no more than trying to uphold and succeed within conventional boundaries, epitomised this. This aspect of Nightingale's thinking shows her already moving towards a holistic conception of health that stressed psychological and moral components—'moral' here defined to encompass 'morale' as well as 'morality'—alongside physiological and sanitary elements.

Once again, however, the harmful health implications of country house life were not the full story. For as subsequent chapters will show, techniques of organisation and management could be transposed from the domestic sphere of the country house onto institutions dedicated to healthcare provision. Nightingale would soon be arguing that making such institutions more homely, in ways recognisably adopted from the models of her experience in her family, would be of great benefit to the nation's health.

Notes

1. Kathryn Gleadle, *The Early Feminists: Radical Unitarians and the Emergence of the Women's Rights Movement, 1831–51* (New York: St Martin's Press, 1995).
2. Nightingale, 'Cassandra', in *The Collected Works of Florence Nightingale, Volume 11: Florence Nightingale's Suggestions for Thought*, ed. Lynn McDonald (Waterloo, ON: Wilfrid Laurier University Press, 2008) [hereafter *CW* 11], 547–592.

3. Ray Strachey, '*The Cause*': *A Short History of the Women's Movement in Great Britain* (London: G. Bell & Sons, 1928), 395–418.
4. Ibid., 29.
5. Evelyn L. Pugh, 'Florence Nightingale and J. S. Mill Debate Women's Rights', *Journal of British Studies* 21, no. 2 (1982): 118–138.
6. Mill to Nightingale, 4 October 1860, cited in ibid., 128.
7. Nightingale, 'Cassandra', 555–556.
8. Ibid., 567.
9. Nightingale, 'Lebenslauf', 24 July 1851, in *The Collected Works of Florence Nightingale, Volume 1: Florence Nightingale: An Introduction to Her Life and Family*, ed. Lynn McDonald (Waterloo, ON: Wilfrid Laurier University Press, 2001) [hereafter *CW* 1], 92; for example, Fanny Nightingale's journal, 4 December 1844, names eleven guests. Claydon N67/4.
10. Nightingale, 'Cassandra', 559.
11. Ibid., 574.
12. Ibid., 573.
13. Nightingale, *Suggestions for Thought*, *CW* 11, 323.
14. Patricia Meyer Spacks, 'The Privacy of the Novel', *NOVEL: A Forum on Fiction* 31, no. 3 (July 1998): 304–316.
15. Nightingale, 'Cassandra', 566–567.
16. Ibid., 567.
17. Ibid., 549.
18. Ibid., 558.
19. Ibid., 561.
20. Pierre Bourdieu, *The Logic of Practice*, trans. Richard Nice (Cambridge: Polity, 1990).
21. Nightingale to Mary Mohl, 16 February 1868, cited in Edward Cook, *The Life of Florence Nightingale*, vol. 2 (London: Macmillan, 1913), 315.
22. Nightingale, 'Cassandra', 549.
23. Ibid., 567.
24. Ibid., 565.
25. Ibid., 574.
26. John Tosh, *A Man's Place: Masculinity and the Middle-Class Home in Victorian England* (New Haven: Yale University Press, 1999), xi.
27. Leonore Davidoff and Catherine Hall, *Family Fortunes: Men and Women of the English Middle Class 1780–1850*, 3rd ed. (London: Routledge, 2019).
28. Tosh, *A Man's Place*, 3.
29. Elaine Showalter, 'Florence Nightingale's Feminist Complaint: Women, Religion, and "Suggestions for Thought"', *Signs* 6, no. 3 (1981): 395–412; Katherine V. Snyder, 'From Novel to Essay: Gender and Revision in Florence Nightingale's "Cassandra"', in *The Politics of the Essay: Feminist Perspectives*, ed. Ruth-Ellen B. Joeres and Elizabeth Mittman (Bloomington: Indiana University Press, 1993), 23–40.

30. Elizabeth Gaskell to Catherine Winkworth, 11–14 October 1854, *The Letters of Mrs Gaskell*, ed. J. A. V. Chapple and Arthur Pollard (Manchester: Manchester University Press, 1966), 308.
31. Regenia Gagnier, *Subjectivities: A History of Self-Representation in Britain, 1832–1920* (Oxford: Oxford University Press, 1991), 33–34. Gagnier utilises Gayatri Spivak's critique of the 'creative imagination' in 'Three Women's Texts and a Critique of Imperialism', *Critical Inquiry* 12, no. 1 (1985): 243–261. For a spatial comparison of 'Cassandra' and Virginia Woolf's *A Room of One's Own* see Wendy Gan, 'Solitude and Community: Virginia Woolf, Spatial Privacy and a Room of One's Own', *Literature & History* 18, no. 1 (2009): 68–80.
32. Nightingale to Harriet Martineau, 30 November 1858, in *The Collected Works of Florence Nightingale, Volume 14: Florence Nightingale: The Crimean War*, ed. Lynn McDonald (Waterloo, ON: Wilfrid Laurier University Press, 2010), 993–995.
33. Nightingale to John Stuart Mill, 12 September 1860, in *The Collected Works of Florence Nightingale, Volume 5: Florence Nightingale on Society and Politics, Philosophy, Science, Education and Literature*, ed. Lynn McDonald (Waterloo, ON: Wilfrid Laurier University Press, 2003) [hereafter *CW* 5], 375–376.
34. Pugh, 'Florence Nightingale and J. S. Mill', 122–123.
35. Nightingale, 'Cassandra', 577, 556.
36. Nightingale, *Suggestions for Thought*, *CW* 11, 508.
37. Monica Feinberg Cohen, *Professional Domesticity in the Victorian Novel: Women, Work and Home* (Cambridge: Cambridge University Press, 1998), 2.
38. Jessica Gerard, *Country House Life: Family and Servants, 1815–1914* (Oxford: Blackwell, 1994), 118; M. Jeanne Peterson, *Family, Love, and Work in the Lives of Victorian Gentlewomen* (Bloomington: Indiana University Press, 1989).
39. Nightingale to Hilary Bonham-Carter, 11 December 1845, cited in Edward Cook, *The Life of Florence Nightingale*, vol. 1 (London: Macmillan, 1913), 44–45.
40. Gillian Gill, *Nightingales: Florence and Her Family* (London: Hodder & Stoughton, 2004), 184.
41. Ibid., 169–189; 1840s letters from Nightingale to Marianne Nicholson, Lotherton Hall, Copeland loan letters, box 3, bundle 1.
42. Adam Kuper, *Incest and Influence: The Private Life of Bourgeois England* (Cambridge, MA: Harvard University Press, 2009), 18; Mary Jean Corbett, 'Cousin Marriage, Then and Now', *Victorian Review* 39, no. 2 (2013): 74–78.
43. Claudia Nelson, *Family Ties in Victorian England* (London: Praeger, 2007), 137; Kuper, *Incest*, 17.

44. Nightingale, 'Cassandra', 581.
45. Gill, *Nightingales*, 283–284; Nightingale to Hilary Bonham-Carter, 25 March 1853, *CW* 1, 446.
46. Nightingale, 'Lebenslauf', *CW* 1, 91.
47. Lynn McDonald, 'Natural Science', *CW* 1, 58–59.
48. Mark Bostridge, *Florence Nightingale: The Woman and Her Legend* (London: Penguin, 2008), 105. Sand's rejection of conventional domesticity, explored in her novels, likely resonated with Nightingale.
49. Cited in Cook, *Life*, vol. 1, 100.
50. Ibid.; Nightingale, 'Cassandra', 571.
51. John Barrell, 'Death on the Nile: Fantasy and the Literature of Tourism 1840–1860', *Essays in Criticism* 41, no. 2 (1991): 97–127; Anthony Sattin, *A Winter on the Nile: Florence Nightingale, Gustave Flaubert and the Temptations of Egypt* (London: Hutchinson, 2010). Nightingale's letters from Egypt were republished by Sattin in 1987, but the reference edition is now *The Collected Works of Florence Nightingale, Volume 4: Florence Nightingale on Mysticism and Eastern Religions*, ed. Gérard Vallée (Waterloo, ON: Wilfrid Laurier University Press, 2002), 117–476.
52. Jill Matus, 'The "Eastern-woman Question": Martineau and Nightingale Visit the Harem', *Nineteenth Century Contexts* 21, no. 1 (1999): 63–87.
53. Nightingale, religious vision recorded in Greece, 1850, in *The Collected Works of Florence Nightingale, Volume 3: Florence Nightingale's Theology: Essays, Letters and Journal Notes Religions*, ed. Lynn McDonald (Waterloo, ON: Wilfrid Laurier University Press, 2002) [hereafter *CW* 3], 225–227.
54. Nightingale to her mother, 12 May 1850, Claydon N121, in *The Collected Works of Florence Nightingale, Volume 7: Florence Nightingale's European Travels*, ed. Lynn McDonald (Waterloo, ON: Wilfrid Laurier University Press, 2004) [hereafter *CW* 7], 397.
55. See *CW* 7, 444–489.
56. Gerard, *Country House Life*, 88.
57. Note by Nightingale, 7 January 1851, *CW* 1, 97–99.
58. Nightingale, note, 1840s, *CW* 1, 94–96.
59. On Parthenope, see also Mark Bostridge, 'Self-sacrifice and Parthenope Nightingale', *Times Literary Supplement*, 24 May 2002.
60. F. P. Verney, *Peasant Proprietors and Other Selected Essays* (London: Longman, Green & Co, 1885); *How the Peasant Owner Lives in Parts of France, Germany, Italy, Russia* (London: Macmillan, 1888); *Memoirs of the Verney Family During the Civil War* (London: Longman, Green & Co, 1894).
61. Adrian Tinniswood, *The Verneys: Love, War and Madness in Seventeenth-Century England* (London: Vintage, 2008); Susan Whyman, *Sociability*

and Power in Late-Stuart England: The Cultural Worlds of the Verneys 1660–1720 (Oxford: Oxford University Press, 1999).
62. Gerard, *Country House Life*, 32.
63. Anne D. Wallace, *Sisters and the English Household: Domesticity and Women's Autonomy in Nineteenth-Century English Literature* (New York: Anthem Press, 2018).
64. Gill, *Nightingales*, 131–132.
65. Richard Monckton Milnes to Viscountess Galway, 1846. UoN MSC, Ga 2 D 1019/1–2.
66. Emily Taylor to Fanny Nightingale, 24 December 1833, Wellcome 9045/4.
67. Nightingale, note, 1840s, *CW* 1, 95.
68. Nightingale to her father, 22 April 1861, *CW* 1, 247–248.
69. Nightingale to her mother, 8 August 1851, *CW* 1, 128.
70. Nightingale to Manning, 15 July 1852, *CW* 3, 253.
71. Nightingale to Manning, 19 August 1852, *CW* 3, 257–258.
72. Nightingale to Manning, 7 September 1852, *CW* 3, 259–260.
73. Nightingale to Manning, 28 September 1852, *CW* 3, 260–261.
74. Ibid.
75. Nightingale, 'Cassandra', 575.
76. Nightingale, *The Institution of Kaiserswerth on the Rhine, for the Practical Training of Deaconesses* (London: London Ragged Colonial Training School, 1851), *CW* 7, 492–511, 493.
77. Nightingale to Manning, 28 September 1852, *CW* 3, 260–261.
78. Nightingale to Manning, undated but following her 28 September 1852 letter, *CW* 3, 265–267. On monomania's origins, see Jan Goldstein, 'Professional Knowledge and Professional Self-Interest: The Rise and Fall of Monomania in 19th-Century France', *International Journal of Law and Psychiatry* 21, no. 4 (1998): 385–396.
79. Elaine Showalter, *The Female Malady: Women, Madness, and English Culture, 1830–1980* (London: Virago, 1987).
80. Sally Shuttleworth, *Charlotte Brontë and Victorian Psychology* (Cambridge: Cambridge University Press, 1996), 34–56, 49.
81. Note by Nightingale, 7 January 1851, *CW* 1, 98.
82. Note by Nightingale, undated [1857], *CW* 5, 233.
83. Nightingale to Manning, 1852, *CW* 3, 265–267, 266.
84. Ibid., 267.
85. Ibid., 266.
86. Selina Bracebridge to Parthenope, undated ['Monday 7', 1853], Wellcome 9045/26.
87. Nightingale mainly used these food parcels 'for the benefit of nurses and colleagues [she] was trying to cultivate'. Lynn McDonald, 'Letters to, from and about Nightingale's immediate family', *CW* 1, 101–102.

88. Note by Nightingale, undated [1857], *CW* 5, 233.

CHAPTER 4

Health at Home

Florence Nightingale's life spanned the period in which Britain developed something approaching a modern public health system. Although there was no Ministry of Health until 1919, by the time of Nightingale's death in 1910 Britain had a General Medical Council regulating the medical profession, a school health service, a Central Midwives Board, expanded teaching hospitals, nurse training schools, an ever-growing network of district nurses and health visitors, workhouse infirmaries providing limited free health care to the poorest members of society, and a whole plethora of underpinning public health laws. Nightingale's life mapped directly onto this first public health revolution. Little to none of this health infrastructure existed in 1820, and most of it came into being during the middle of the century and after the Crimean War. Nightingale undoubtedly played an important role in shaping the form and speed of some of these transformations. Her popular advice, on everything from the proper way to air a room and dispose of excrement to the safest material with which to cover walls, helped publicise the wide-ranging influence of physical environments upon human health. Although the wider growth in public health would have happened without her, she should be situated as contributing to this age of unprecedented medical progress and infrastructural development. The appetite for betterment in an age of industrialisation was ultimately unstoppable.

To a significant degree, both the discursive rhetoric and practical plans to transform the public health of nineteenth-century Britain centred on

P. Crawford et al., *Florence Nightingale at Home*,
https://doi.org/10.1007/978-3-030-46534-6_4

the home. Maintaining clean, healthy, respectable homes became the royal road to promoting physical and mental (or moral) health. This contrasted with the second half of the previous century, in which efforts to improve health had instead been focused on public spaces. Between 1760 and 1799 there were significant government-driven enhancements to the lighting, paving, and sweeping of public areas, yet, as Karen Harvey has argued, there was an 'absence of the household from eighteenth-century theories of political obligation'.[1] By the 1850s, however, the home had become central to both political and wider cultural discussion, including that surrounding health.

To prevent the emergence and spread of disease, it was argued, homes needed to meet higher standards of cleanliness and sanitation. Germ theory was not yet established, but it was frequently noted that dark, dank, dirty, ill-ventilated spaces were the ones in which diseases festered. Consequently, the key to improvement was to drive the fetid and unsanitary away from domestic spaces, both through modifications in people's behaviours and by new sanitary technologies. An early flush lavatory, for example, constituted a key attraction in the famous 1851 Great Exhibition in London showcasing international design and manufacturing. Other celebrated technological improvements were larger in scale—most notably the construction of municipal sewer systems, to which all new houses had to be connected from 1848. As much as these new technologies of sanitation influenced town planning, they also, crucially, influenced the design of, and habits enacted within, domestic spaces.[2] Home was where good, healthy habits started, the springboard from which society would become stronger and more productive.

This message created a tension. The Englishman's home might have been popularly supposed an inviolable castle, but this new vision of public salubrity required regulation and inspection. Even in their homes, the population—by which, in practice, legislators and reformers tended to mean the working-class population—needed to be subject to an outside gaze. In Tom Crook's words, the populace was 'counted and classified according to norms of good health; and, ultimately, posited like an *object* governed by scientific laws of health'.[3] As this framing suggests, the inherent tension presented through this new enthusiasm for state intervention into the domestic sphere centred on class: it was middle- and upper-class reformers who went into working-class homes and sought to change the principles by which their occupants lived. Over the course of the nineteenth century, various kinds of visitor crossed the threshold into workers' cottages, concerned with the moral and physical health (often in that order) of their inhabitants. Many of these visits were the result of private arrangements or part of philanthropic expectations, but as the century progressed the visits became discernibly more organised, bureaucratised, and centralised. They included upper-class charitable visitors, the 'Ladies Bountiful' of the landed classes who practised philanthropy on their estates as part of maintaining, as Jessica Gerard writes, 'the personal contacts so crucial for maintaining the system of patriarchal control and deference'.[4] Nightingale's family introduced her to this tradition during her youth. Organisations such as the Ladies' Sanitary Association, founded in 1857 with branches in nearly a dozen cities by 1865, were devoted to educating working-class mothers in domestic management, distributing everything from soap to instructive tracts to thousands of homes. Members of the clergy and religious missionaries also joined the visiting tribes, especially after the 1851 census revealed a startling lack of engagement with religion among the labouring classes. Nightingale's experiences very much reflected this broader trend.

In the middle of the century, home visiting extended to prominent novelists, journalists, and artists who, as Carolyn Steedman evocatively describes, sought to 'penetrate the maze of greasy streets and step through the door of another kind of space'.[5] The domestic scene became an evocative subject, with writers and artists all contributing and responding to a cultural fascination with the simultaneously familiar and dangerous associations of the domestic space. Written accounts of life in slum housing, in particular, exposed readers to a set of social and economic circumstances of which they were often unaware.[6] In the 1840s, this was equally true of a reformist liberal like the sanitarian Edwin Chadwick as it was of the revolutionary socialist Friedrich Engels. The social researcher Henry Mayhew, whose 1851 *London Labour and the London Poor* provided a vivid sketch of the capital's underclass, was another such writer-visitor. In an 1853 essay 'Home Is Home, Be It Never So Homely', Mayhew contrasted a middle-class home of 'comfort and affection' with a house rendered 'no home at all', the latter suffering from its 'leaky roof, broken windows, damp walks, reeking drains, and wet clothes hung to dry'.[7] The homes of the poor had become sites of enquiry, appearing to invite both dramatic description and sanitarian intervention from those further up the social hierarchy.

Nightingale was part of this surge of concern and change. The conceptions of sanitary homes that she put forward in the second half of the century were rooted in the practice of charity that she had regularly undertaken with female relatives in the 1830s and 1840s, notably in the area around the family's summer home of Lea Hurst, Derbyshire. In that house, the Nightingales were situated much closer to the local working-class community than was typical for people of their social position, or compared to at their grander house at Embley, where, as Nightingale explained, the 'village population was very much scattered and the park so large that no cottages were very near'.[8] In Derbyshire, by contrast, the presence of Lea Mills gave the family close and daily exposure to a large number of textile factory workers and their dependents, in whom Nightingale took considerable interest, forming relationships and a sense of quasi-pastoral responsibility that endured into later decades.

In the pages that follow we will first examine the Nightingale family's practice of charitable health visiting as part of a wider pattern of responsibility placed on upper-class women. We will then consider Nightingale's writing on health in the home, especially her bestselling book, *Notes on Nursing* (1860), which encouraged ordinary people, but particularly women, to take responsibility for managing healthy homes. Nightingale's fascination with visiting, describing, and improving interior environments brought her into contact with other figures in the public health movement who also understood the importance of advising on domestic arrangements and habits for illness prevention. The final part of the chapter will look at her contribution from the 1860s onwards to the development of district nursing and health visiting alongside philanthropists such as William Rathbone. These schemes were the beginnings of a formal system, representing a more secular and institutionalised version of the aristocratic tradition of home visiting to which Nightingale had been introduced in her youth. Nightingale drew on her society's fascination with private spaces and made homes the starting point for much of her programme of health reform.

Family Practices of Home Visiting

In 1891, Nightingale expressed frustration to her nephew Frederick Verney regarding the British population's response to issues affecting their everyday health. Despite prompting wide-ranging reforms, the 'flood of sanitary books, pamphlets, publications and lectures of all sorts' had, in her view, barely altered the everyday habits of most families. 'What is read in a book stays in the book', she complained to Verney, 'Health in the *Home* has not been carried *home* to the hundreds of thousands of rural mothers and girls upon whom so largely depends the health of the rural population'. In what turned out to be something of a final attempt to address this general lack of awareness, Nightingale proposed a new scheme to spread health advice to poor families in which educated women would 'personally bring their knowledge home to the cottager wives'. Central to this 'mission of health for rural districts' was the physical interaction between poor rural families and a newly trained cohort of what she called health missioners. These women 'must be *in touch* and in love, so to speak, with the rural poor mothers and girls and know how to show them better things without giving offence'.[9] When Nightingale presented a more developed version of the scheme to the Conference of Women Workers in 1894, she again emphasised the importance of this close and personal contact for overturning the 'habits of dirt and neglect' in poor cottages. Health missioners needed to cultivate 'sympathy' for the poor, which, Nightingale explained, 'can only be got by long and close intercourse with each in her own house'.[10]

Despite its apparent novelty, this 1890s scheme recalled a far older tradition of care for the poor with which Nightingale was personally acquainted. By endowing 'women of the highest cultivation and of the deepest sympathy' with special responsibility for the cleanliness of rural cottages, and thereby the physical health of the working population, Nightingale sought to incorporate the long-established aristocratic practice of visiting the homes of the sick and the elderly to which she had been exposed in her adolescent years into the public realm of late nineteenth-century health and social work. Organised around 'direct contact between the donor and the beneficiary', such visits enabled a family of the Nightingales' standing to offer a degree of care to the communities of workers and tenants for whom they felt responsibility.[11] Nightingale's 1894 call for health missioners showed how integral the private sphere of the home remained in her thinking. 'In all European countries, more sickness, poverty, mortality and crime is due to the state of our poor men's dwellings than to any other cause', she wrote in the early 1860s, explaining that 'I would rather devote money to remedying this than to any institution'.[12]

By the age of ten, Nightingale was already showing concern for the often shocking domestic trials of the poorer members of the communities around her, noting in a diary entry of 1830 that 'on the 29th of May at Brooke Lodge, Mrs. Petty killed herself with her youngest child at five o'clock in the morning'.[13] In her diary, Nightingale recalled the scene as she had heard it described by the adults around her, including the detail of Petty's husband attempting to beat down the door in order to reach his wife in time. The episode showed Nightingale's early willingness to confront the more difficult aspects of life in her (generally very privileged) surroundings.

Nightingale drew further connections between domestic cleanliness and physical health when she was encouraged to meet and interact with the labouring-class communities living near her family homes. It fell on the women of the Nightingale family to manage the philanthropic aspects of the relationship with these groups, especially during the first decades of the nineteenth century, when involvement by upper-class women in charitable endeavours was growing.[14] By 1830, women such as Nightingale's mother Fanny were carrying out hundreds, even thousands, of district visits annually—mostly to poor people's homes, but also encompassing community work in schools, hospitals, chapels, lying-in institutions, and soup kitchens. Such work could be done without compromising contemporary views of women's virtues and roles, and indeed was supported by proponents of the philosophy of the separate spheres as a fitting means of female occupation. In *Coelebs in Search of a Wife* (1809), for instance, the Evangelical writer Hannah More encouraged 'the affluent to [come to] a nearer knowledge of the persons and characters of their indigent neighbours'.[15] Visiting the poor in their own homes exposed landowning families to the interiors of cottages on their estate and the variety of illnesses found within them. However, crucially, it also underscored and reinforced class differences, by aligning the alms-giving wealthy with the respected social role of educator and philanthropist, while placing the poor in the role of deferential, grateful recipient of material assistance and spiritual guidance. Furthermore, texts such as Christopher Benson's *District Visitor's Manual* (1840) celebrated the 'social, moral, spiritual benefit' of this regular exercise of religious and paternalist duty to the visitors themselves.[16] Acts of charity may have been undertaken sincerely and for their own sake, but they were also a performative enactment of social status that satisfied the ladies' own needs for purpose and vocation while elevating their self-esteem and the regard in which they were held within community hierarchies.[17] For these reasons, Nightingale was to later criticise such 'odds and ends of charity' as a 'kind of conscience quieter, a soothing syrup'.[18]

Correspondence of the 1830s between Fanny Nightingale, her aunt Maria Coape, and sister Julia Smith shows how the family participated in this upper-class enthusiasm for visiting. 'The sick list has been rather formidable', began Coape in one letter to Fanny Nightingale of March 1834. Physical assistance was often given to those with less serious illnesses who showed the promise of returning to work, with Coape recording how a 'box of pills' that she had provided enabled one villager to return to work 'at the mill again'.[19] In another letter Smith shared observations of the cottagers with particularly severe or long-lasting conditions. This included a woman in Lea who 'had a stroke about three weeks ago after which she never left her bed and died in about ten days', as well as a Mrs Holmes who 'died quite resigned and without pain, early on Saturday morning'. Smith's account included the young Nightingale as a part of the group of visitors spending 'a considerable time' with the terminally ill: 'we talked much about you all but more particularly Flo, for whom she [the ill person] offered up a silent prayer'.[20] The large range of names that crop up in such letters, and the presumption of a pre-existing familiarity with each villager's condition on the part of the recipient, demonstrate the sustained community engagement on the part of the Nightingale women.

Unlike his female relations, William Nightingale paid little more than passing attention to the health-related repercussions of poorly designed habitations. He interacted with villagers not in the capacity of providing charity, but as a landlord and—in his mind's eye—a Romantic traveller. While on 'perambulations thro' [*sic*] the many tenements' of his estate in the 1820s, he described the 'walls and roofs and doors and windows' of the cottages as his 'to-do or not-to-do'. Accounts of illness did occasionally crop up in these descriptions, but only as incidental background, not as a problem with which William actively engaged on a personal level. He described the cottage interiors in figurative, not practical, terms:

> The stirring little dairy-farmer preparing his fat pig for the slaughter, with wife of tidiness and floors and dress of indescribable neatness. Then again the house too dirty or too unsightly to admit one across the threshold with women unlike humanity. Oh! The contrasts of human things. Beauty and ugliness – roughness and softness – the attractive and the repulsive. What an infinite society is here, if every house was but a little more neat and nice than it is – and every man, woman and child a little cleaner and more picturesque in outward show.[21]

In contrast to the personalised observations made by the Nightingale women, William Nightingale's notes did not refer to the names of the cottagers. It was not his business to concern himself with the vicissitudes of life behind the closed doors of the dwellings. Coape recalled that '[a]t every turn I hear the people saying "Mr Nightingale was in such a hurry that I could not speak a word to him"'.[22] His detached observations were also indicative of broader changes in relieving poverty at state level in which he was politically involved—he sat on New Poor Law Boards of Guardians in both Derbyshire and Hampshire. As Anne Summers has argued, over time these male-dominated, state-affiliated replacements to the longstanding system of parish-based outdoor relief 'deprived the lady visitor of much of her social power and influence'.[23]

As a young woman, Florence Nightingale was therefore introduced to two quite different approaches to poor relief. On the one hand, the new, utilitarian system associated with her father's politics informed her subsequent contributions to more systematic, efficient approaches to human care, even if she disliked the reigning orthodoxies of political economy.[24] On the other hand, her female relations' visits to those on the sick list, and the tradition of aristocratic charity of which this was a part, provided an awareness of illness in individual working-class homes that was to re-emerge time and again in her later writings and personalised approach to training district nurses and health missioners.

By the age of sixteen Nightingale was treading a well-worn path into the homes of the needy and ill. In 1836 Parthenope wrote to her grandmother to inform her that 'Flo has been very busy paying visits in the village', remarking with some pride that the locals were 'very fond of her' and that Nightingale 'likes them and is always sorry to leave them'.[25] Nightingale's experience of disease was not always confined to these working-class areas, however. For instance, a January 1837 outbreak of influenza provided a particularly notable opportunity for Nightingale to take an active role on the family's Hampshire estate in administering care, finding herself nursing her family and fifteen servants back to health with only her cook to assist.[26] While at the Deaconesses' Institute in Kaiserswerth in 1851 (Chapter 5), Nightingale looked back at the 1837 epidemic almost fondly. Although the outbreak had been distressing and worrying, it had given her an opportunity to act practically and exceptionally in what she otherwise deemed to be a 'wholly unpractical' early life.[27]

Nightingale's own home was therefore also a place in which she was exposed to her later vocation: along with the local working-class houses she visited, it was the test bed for her nursing. She welcomed 'anything' that associated her with 'any class not my own' and, as we have seen, enjoyed activities that took her outside the bounds of upper-class family routine.[28] In a letter to her sister during the 1837 outbreak, Nightingale described how she comfortably assumed the duties of 'nurse, governess, assistant curate and doctor'.[29] Wholly fitting of her character, however, her gaze also turned outward to the more serious effects that influenza was having in the surrounding parish at Embley, where working families were struggling against 'one mass of illness'. Though epidemics affected families and homes of all standings and sizes and forced the population to consider the collective dimension of their lives, factors such as diet, damp surroundings, and the high cost of fuel meant that influenza was far more deadly in the conditions in which the poor were living.[30] For Nightingale, the 1837 influenza epidemic widened the scale of what may have otherwise remained a parochial understanding of living conditions in the cottages. It was closely associated with her famous call to service the same year, when she reported God commanding her to expand her work to save lives.

When Elizabeth Gaskell visited Lea Hurst in 1854, she was told an evocative story in which the young Nightingale disappeared into the surrounding villages at night-time, only to later be found by her mother sitting with a lantern 'by the bedside of someone who was ill'. Nightingale refused to return home, insisting she could not sit down to 'a grand 7 o'clock dinner' while such suffering was taking place.[31] This striking image of the young Nightingale, huddled beside a sick cottager, shows how (in family folklore at least) she found her own place—her own home—in the role of carer for the less socially and economically advantaged. It was not the making of such visits that made Nightingale unusual—as we have seen, this was a conventional activity for a woman of her background. It is notable, however, that she seemed to treat this aspect of an upper-class lady's social role with an unusual degree of seriousness and sincerity. There was nothing performative about Nightingale's visits. On 16 July 1846 she noted in her diary that '[a]ll the people I see are eaten up with care or poverty or disease. When I go into a cottage I long to stop there all day, to wash the children, relieve the mother, stay by the sick one. And behold there are a hundred other families unhappy within half a mile'.[32] Furthermore, Nightingale's commitment to those living around her Derbyshire home lasted for decades after she had physically left the area. In 1879 she was spending around £500 on what she called 'the old and sick women' of the community around Lea Hurst, employing a doctor who would visit the sick at home and update her on their condition.[33] Nightingale even visited some cottages herself during a spell at Lea Hurst in 1881, when she recorded putting more than sixteen afternoons aside to 'give to the village people'.[34]

But it was in the middle of the century, as the powerful forces of industrialisation and urbanisation made Britain's social divide more acute, that the home visit took on a wider symbolism in terms of social healing. Gaskell's novel *Ruth* (1853), like Harriet Martineau's *Sickness and Health of the People* (1853) and later George Eliot's *Romola* (1863), featured a Nightingale-like protagonist visiting the homes of the victims of an epidemic. The female characters in such 'outbreak narratives' sought to heal not only the patients but also the divides and indifference felt between different social groups.[35] The anecdotes that Gaskell heard about the young Nightingale probably also influenced the final chapters of *North and South* (1855), which Gaskell wrote during her stay at Lea Hurst. The novel reflected mid-century debates about the role of women in the public sphere, as well as the related question of how the poor were to be reached and introduced to supposedly proper values through philanthropic social intervention. How and when to enter homes under the auspices of improving health and offering comfort was a question that Nightingale was to return to repeatedly in her post-Crimea career.

Notes on Nursing

Nightingale began her most significant intervention into the houses of ordinary people in the late 1850s, following the partial success of her work for the Royal Commission on the health of the British army in 1857–1858 (Chapter 7). By 1859, Nightingale felt that the momentum for far-reaching sanitary change was stalling. Not only was the Chief Medical Officer, Sir John Simon, relatively unsympathetic to her ideas, but her political champion, Sidney Herbert, was terminally ill with pleurisy, and Alexis Soyer, the campaigning chef and Crimean War comrade, had died. Nightingale had consequently fallen into what her cousin Beatrice Smith described as a 'state of extreme discouragement'.[36] To remedy this despair, Smith introduced Nightingale to Edwin Chadwick, the leading sanitary reformer of the age. Their subsequent exchanges not only energised Nightingale with a new purpose, but also prompted one of the most distinctive health campaigns that the country had seen.

In the 1850s, Nightingale and Chadwick both subscribed to miasmic, anticontagionist theories of disease.[37] According to this then widely accepted framework, disease was caused by filth and its odours rather than specific pathogens, and its spread could be curbed by the removal of decaying matter and the implementation of good sanitary principles. It is a 'continual mistake', wrote Nightingale in 1860, 'to look upon diseases, as we do now, as separate things which must exist, like cats and dogs'. Instead, she continued, diseases should be thought of 'as conditions, like a dirty and a clean condition, and just as much under our own control'.[38] For Nightingale, a non-medically trained sanitarian, anticontagionism explained the various diseases she encountered in the homes of the poor while enabling her to retain the strongly ethical outlook of sound moral management that was central to her religious views. It protected the moral dimension of human conduct and reinforced, in Charles Rosenberg's words, 'the traditional assumption that both health and disease arose from particular states of moral and social order'.[39]

Chadwick had been very successful at translating these anticontagionist theories into action on the ground. From the 1840s onwards he had influenced a dramatic shift in popular understanding of sanitation, most famously in his four-hundred page *Report on the Sanitary Conditions of the Labouring Classes of Britain* (1842) that linked unsanitary living to the prevalence of disease among the poor. Setting new standards of rigour and breadth, the report based its depiction of the population sleeping in cellars amidst horses and pigs, with no disposal place for human waste and poor drainage allowing sewage to fester on the street, on the testimonies of around 2000 Poor Law Commissioners, medical officers, factory inspectors, and other experts. While Nightingale had first-hand experience of such conditions, the report brought the grim circumstances of the poor to the attention of others from similarly comfortable backgrounds, prompting, as Nancy Metz has put it, 'the middle class to discover the poor'.[40] Crucially, Chadwick's report led to the 1848 Public Health Act, which introduced new developments in drainage, water supply, sewage treatment, and road construction in towns and cities across the country. The Act has been understood in different historical accounts as either a progressive crusade against disease or a conceited effort to control an unruly population and divert attention from the underlying causes of poverty.[41] Either way, it marked a turning point after which 'government intervention into matters of health and sanitation [would] seem not only acceptable but inescapable'.[42]

Chadwick carried both the successes and the frustrations of sanitary reform into his exchanges with Nightingale. His earlier influence over government policy was reduced when in 1854 he was ousted from the Board of Health by the less sanitary-minded John Simon. In Nightingale, and the enormous prestige and popular acclaim she had earned during the Crimean War, Chadwick therefore saw an opportunity to effectively go over the heads of the politicians by bringing 'prevention within the means of popular appreciation', as he wrote to her.[43] Nightingale had hitherto focused on petitioning those in positions of influence to bring about changes in policy: the eye-grabbing polar area chart that appeared in her *Notes on Matters Affecting the Health of the British Army* (1858), for example, had instigated discussions of soldier mortality rates and disease at the highest levels of government.[44] Chadwick reminded Nightingale, however, that such a 'great monogram of evidence' would be 'shut out from the people' if written 'only to a select class, and on a professional subject'. Instead, he urged her to consider a 'change in direction of your labour' by suggesting, in what was to become *Notes on Nursing*, 'advice to nursing mothers or to young mothers on sanitary treatment before the arrival of the physician'. Alongside Nightingale's post-Crimea public image as 'the great national nurse', her 'clearness and commensurate force of exposition' would 'be heard by hundreds of thousands and millions, who will give a deep and well-deserved attention'.[45] Chadwick's suggestion was part of a general shift among mid-century sanitary reformers to prioritise changes to popular behaviour over the past focus on infrastructure and law. Such top-down changes would 'only result in partial reforms', argued the women's rights campaigner Bessie Rayner Parkes at the 1859 meeting of the National Association for the Promotion of Social Science (NAPSS), 'until the habits of the people, engendered amidst bad conditions, and rendered careless by hopelessness, be also changed'.[46]

Following Chadwick's suggestion, Nightingale set to work on *Notes on Nursing* (1860), a book that explained in meticulous detail how inhabitants themselves could prevent illness by overseeing minor domestic improvements and adaptations of behaviour. It placed women at the very centre of sanitary reform—correcting their somewhat peripheral role in Chadwick's *Report*, which was more concerned with the domestication and depoliticisation of working-class men.[47] Typical of Nightingale's more general emphasis on what Rosenberg terms the 'role of volition and behaviour in the causation of infection', *Notes on Nursing* called on all readers, regardless of their class origins, to shape their own health and prosperity.[48] What marked Nightingale's text out from the other forms of sanitarian education to emerge at this time was its accessible style and its female authorship. *Notes on Nursing* was among the first text of its kind to address women in their own homes and, as a popular domestic handbook, offer accessible advice on the management of domestic sickrooms and general practices of hygiene.

Nightingale was aware that *Notes on Nursing* had to avoid seeming patronising in order to reach the varied audience that she desired. She recognised that much writing on sanitation drew on wider paternalistic attitudes, wryly noting in a journal entry of the 1840s how people of her class tended to 'acknowledge but two virtues in "poor people" – cleanliness and gratitude'.[49] From its opening lines, *Notes on Nursing* signalled its intention to instead address the individual in terms of their responsibilities for their home. Contrary to what many modern readers may assume, the book was not intended 'as a manual to teach nurses to nurse', but instead, as Nightingale wrote, as 'hints' for those 'who have personal charge of the health of others'.[50] The *Quarterly Review* wrote of the book that 'the nurse spoken of is not usually the hired professional person, but the wife, mother, or sister'.[51] To distance *Notes on Nursing* from what Patricia Branca has called the 'naggedly critical' tone of most household manuals,[52] Nightingale explained to her (presumed female) reader that she was not attempting 'to teach her', but instead asking 'her to teach herself'.[53] By beginning the text in this way, Nightingale situated her work within a parallel genre of self-improvement writing that aimed to aid the aspirational reader in their journey towards individual progress, exemplified the previous year by the Scottish author Samuel Smiles's widely read *Self-Help* (1859), famous for its maxim 'heaven helps those who help themselves'.[54] Within the wide range of responsibilities to which the doctrine of self-help could be applied, *Notes on Nursing* called upon the individual mother—the guardian of her home—to control her indoor environment for the sake of the health of her family. Nightingale's emphasis on the capacities of the morally responsible female cottager was echoed by Smiles's later work *Character* (1871), in which he informed his readers that 'even the poorest dwelling' could become 'the abode of comfort and virtue and happiness' when 'presided over by a virtuous … woman'.[55]

If Nightingale made strenuous efforts to avoid being patronising, she was also aware of the risk that *Notes on Nursing* would seem too dogmatic, or even boring. In a letter of February 1859, her close advisor Dr John Sutherland warned her that 'the public did not appear to relish' attempts to educate them on domestic sanitation, advising that her book should be 'more preceptive, and less doctrinal'.[56] Part of Nightingale's attempt to do something new came through her personalising the book with regular allusions to her own experiences and public reputation. She presented herself as an 'experienced observer' familiar with 'watching disease', but also tacitly recalled her Crimean experience by stating that 'I myself have had sorrowful experience, from open sewers loaded with filth'.[57] When writing to Martineau, lauded for her accessible writing style in *Illustrations of Political Economy* (1832–1834), Nightingale explained that she thought 'the only possible merit of my little book is that there is not a word in it, written for the sake of writing, but only forced out of me by much experience in human suffering'.[58]

In fact, *Notes on Nursing* featured far more rhetorical skill and persuasive technique than this modest claim of a 'little book' would suggest. With phrases such as '[a] dark house is always an unhealthy house, always an ill-aired house, always a dirty house', Nightingale made stylish use of repetition, drawing on Sutherland's advice to 'be simpler' and to include a 'few easy sentences requiring little thought'.[59] Nightingale coined short, memorable phrases, such as 'air within as pure as the air without', which she repeated for emphasis, and then confronted her readers with urgent questions, asking, provocatively, if it was better 'to learn the piano-forte than to learn the laws which subserve the preservation of offspring?' Allusions to death appeared unexpectedly in disarming places, for instance when Nightingale invited her readers to imagine a murderer cutting 'the throat of a poor consumptive creature, sitting by the fire' and to then compare this to the equivalent threat of death lurking 'in the musty, unaired, unsunned room', including 'the scarlet fever which is behind the door, or the fever and hospital gangrene which are stalking among the crowded beds of a hospital ward'.[60] *Notes on Nursing* was more than a straightforward reproduction of Nightingale's 'experience in human suffering': this multifaceted, often dramatic narrative combined moral advice with medical statistics and grisly anecdote in inventive ways.[61]

Beyond its stylistic prowess, *Notes on Nursing* provided its readers with clear and useful advice, including tips on the position of windows and chimneys in relation to that of beds and the arrangement of clothing racks for drying damp towels in bedrooms. Opening windows and letting 'air without' circulate featured prominently—anything that made the air 'stagnant, musty and corrupt' was, according to Nightingale's anticontagionist worldview, 'ripe to breed small-pox, scarlet fever, diphtheria, or anything else you please'.[62] Once these causes of disease were brought to life on the page, Nightingale hoped that the illness in readers' homes would be swept, dusted, and scrubbed away.

Nightingale's publisher, Harrison, released over 15,000 copies of the first edition of *Notes on Nursing* in the first two months.[63] The *Quarterly Review* lauded it as a 'genuine operation of genius' and applauded its 'stimulating style' that allowed for an 'amount of meaning [to be] conveyed in the shortest and sharpest way'.[64] *The Times* similarly admired its 'pointed and pithy terms' and 'force of brevity',[65] while Chadwick, reading the final version of a work that he had suggested, expected it to 'circulate wider and permeate deeper' than other texts and to 'lead to extensive voluntary action, in [a] direction which may not be foreseen'. He suggested to Nightingale that a future edition should be reduced to a price below two shillings and be 'more widely spread than the publisher may be accustomed to spread his publications'.[66] The issue of the book's price raised questions about the wider value of such writing on health. Was this essentially a text for the amusement of middle-class readers priced according to the norms of the marketplace? Or did the value of the book in advancing the health of all—a 'national service', as the *Daily News and Express* put it—make any discussion of price irrelevant?[67] *The Examiner*, at least, judged *Notes on Nursing* to be 'an inexpensive pamphlet, not costing the price of one bottle of medicine', and applauded its preventative advice for reducing 'the doctor's bill in many families'.[68]

These considerations of domestic economy prompted a new, ninety-six-page edition in April 1861 at the reduced price of sixpence, with the revised title, *Notes on Nursing for the Labouring Classes*. Nightingale's textual amendments for this cheaper edition demonstrated her increasing concern to render homely details in ways that would resonate with a working-class readership. To aid accessibility, she simplified descriptions of home environments to exclude references to upper-class domestic refinements. For example, an allusion to walking along fashionable Regent Street was rephrased as moving, simply, 'up the street'. A 'mother languid' who in the first edition could only travel beyond her home by carriage was limited 'to her house' in the revised version, while 'foul air' was now to be found in 'private houses' rather than 'handsome private houses'.[69]

Two pages of instructions on methods for introducing fresh air into small country cottages were added, as were two new chapters, 'Health of Houses' and 'Minding Baby', inspired, as Nightingale explained to her mother, by the 'Lea Hurst cottages' on her family's land.[70] For the critic Viktor Skretkowicz, such changes diminished the 'mesmerizing effect of Nightingale's fascinating prose style', which is to say that the burden of a diverse readership compromised the rhetorical effect of the earlier library edition.[71] But this interpretation disregards the perceptive way that Nightingale constructed a more inventive form of writing and found new ways to communicate with, and ultimately influence, her new audience.

Nightingale's revisions to her domestic health advice manual were therefore borne of her familial experiences and past encounters in cottages as well as her more recent anticontagionist views. When the new edition landed in the hands of the statistician and epidemiologist William Farr, he announced it to be '*dearer* than ever to the friends of sanitary reform'.[72] *Notes on Nursing* showcased Nightingale's unwavering conviction that the home should be the nucleus of reform, demonstrated her faith in the management skills of the population, and cannily tapped into the mid-century enthusiasm for self-help. Crucially, it suggested that the overall effectiveness of high-level sanitary reforms depended on complementary forms of popular health communication.

District Nursing: The Return of the Lady Visitor?

The sanitarian model of disease prevention that Nightingale endorsed from the 1850s onwards introduced the possibility that a far wider range of people could access knowledge to support their health. Print technologies provided a ready means of bringing such knowledge into the household. Following *Notes on Nursing*, such works as Osborne Reynolds' *Sewer Gas, and How to Keep It Out of Houses: A Handbook on House Drainage* (1872) taught women at home how to accommodate and adapt to what the historian Michelle Allen has called the 'altered sanitary landscape'.[73] Rather than awaiting the occasional proprietorial visit of their apparent superiors, with the assistance of these books readers could help themselves to become healthy.

However, the limitations of this health literature simultaneously prepared the ground for the lady visitor's return in physical form. Despite the changes that Nightingale made to the cheaper edition of *Notes on Nursing*, the sanitary insights located within its pages would have remained inaccessible to many working people. Its success was likely due to the enthusiasms of an educated, middle-class readership whose homes, and indeed health, were in less urgent need of improvement. Furthermore, domestic health manuals were often unreliable. Nightingale pointed out that books such as Catherine Buckton's *Health in the House* (1875), *Our Dwellings, Healthy and Unhealthy* (1885), and *Comfort and Cleanliness* (1894), which included clear diagrams and were aimed at semi-literate women, exhibited 'bad mistakes in science'.[74] Finally, there was the question of whether a book—even if properly researched and read by the right audience—would change behaviour and improve the quality of ordinary lives. At the NAPSS meeting in 1859 Parkes had cautioned against merely reading about sanitary matters and instead underlined the 'need to inspire' woman householders 'with a living horror of ill health'. Parkes recognised that what she called 'the sanitary idea' needed to be enlivened with the enthusiasm and energy of entertainment, to become 'a branch of popular knowledge, and a source of popular and enthusiastic activity'. Ever interventionist, the Victorians had a ready-made model for overcoming public apathy and injecting seemingly dour reading material with popular enthusiasm. Recalling the recent movement to counter 'religious indifference', Parkes admired how the 'Bibles and tracts broadcast over the country' had translated into 'Sunday classes and ragged schools, in new churches, city missions, [and] mother's meetings'.[75] Using this as a precedent, the Ladies' Sanitary Association distributed its sanitary tracts via well-established networks of bible-women: a move then parodied in an 1872 cartoon in *Punch* entitled 'Sanitary Sermons'.[76]

Prevailing religious and community structures were drawn into this new 'cult of cleanliness' that 'carried', in the words of the public health historian Antony Wohl, 'with religious fervour down to the masses'.[77] Upper-class families, charitable organisations, and the Church were still able to harness enthusiasm among local communities, and they thereby offered channels through which sanitarian messages could effectively penetrate the otherwise unreachable private sanctuary of the home. 'In those million centres we call the home', argued Dr Benjamin Ward Richardson in his 1880 essay 'Woman as a Sanitary Reformer', 'sanitary science must have its true birth'.[78] It is within this context that sanitarians and the utilitarian bureaucracy found a new use for a charitable, Lady Bountiful-like figure. As Parkes had announced in 1859, the aim was to unite everyone from 'the peeress whose husband owns half the county' to 'the district visitor who cares for the soul' in the cause of health: 'we want the action *of women* in every parish; we want the clergyman's wife and the doctor's daughter to know the laws of health, and to enforce them in the perpetual intercourse which we hope and believe they maintain with their neighbours'.[79]

Perceived as innate carriers of sympathy, able to communicate to all classes, and occupying a social position that obliged them to interact with the poor, women of the middle and upper classes were already engaged in 'perpetual intercourse' with their neighbours and therefore ideally placed, as Parkes explained, to ensure that sanitary knowledge was 'worked into the public mind' and 'condensed in domestic conversation'.[80] The medical periodical *The Lancet* announced its rediscovery of this age-old role in similar terms when it lauded the upper-class lady for her special access to the intimacy of 'the mechanics' room and the poor man's cottages', working in contrast to the health boards that had been 'rather of a public than a fireside nature'.[81] Octavia Hill played a vital role in this revival of the lady visitor in the public imagination through her leadership of the Charity Organisation Society and her public declarations celebrating the virtues of encountering the poor in their homes on familiar turf, advising prospective volunteers in 1877 that '[y]ou might teach and refine them and make them cleaner by merely going among them'.[82]

Nightingale was herself a well-known example of a lady with the ability to communicate with popular audiences and communities on matters of health. In 1855 the physician Edward Henry Sieveking was among the first to suggest that the personable, intimate image of the lady with the lamp could be adapted to help popularise sanitary guidelines (Chapter 6).[83] If the supposedly masculine activity at the War Office could benefit from such 'co-operation of the female sex', Sieveking told the women in his audience that they could similarly bring what he called 'the knowledge' of male science and medicine to 'bear beyond your homes'.[84] The effect of this rhetoric was evident in 1859 when *The Philanthropist* lauded the 'glorious example set by that angelic woman [Nightingale]' for causing 'ladies of elevated social position [to] seem anxious to ameliorate the condition of their poorer sisters'. The Crimean War had provided a visible illustration of the meaningful roles that women could go on to pursue in peacetime and, although there were no longer any 'wounded warriors to tend', the columnist reminded readers that they could imitate Nightingale by giving 'good counsels … to the impoverished and the ignorant'.[85]

This shifting meaning of womanhood also helped Nightingale to supplement and extend her previous efforts in health writing. Her next public health scheme was district nursing: a new route for women to systemically enter the poor person's cottage as carer, and, to some extent, expert advisor. Alongside the Liverpool industrialist William Rathbone, Nightingale played a pioneering role in the development of this new profession.[86] Rathbone had contacted Nightingale in 1861 for help extending to the Liverpool poor the kind of private nursing care that was available to the wealthy, and whose benefit he had seen in his wife on her deathbed two years earlier. The urban poor otherwise had few healthcare options in their homes. Those who avoided the workhouse infirmaries and who were unable to access voluntary hospitals (which often only treated restricted sets of patients) might enlist the service of a domiciliary nurse to care for them at their house. But domiciliary nurses had a poor reputation and little training.[87] In responding to Rathbone's proposal, Nightingale had emphasised that, in contrast, the new figure of the district nurse would not undertake 'amateur work', enclosing a character sheet drawn from the appendix to *Notes on Nursing* to demonstrate what she deemed to be a nurse's essential qualities.[88] Even though this earlier text addressed ordinary women in their homes rather than those explicitly working as nurses, Nightingale was now happy to direct Rathbone to the section titled 'What is a Nurse?' for guidance on developing the profession. Judging district nursing to be the most important of all branches of nursing, Nightingale insisted, in Rathbone's 1865 account, that the 'powers and influences in sanitary matters' required of those treating the poor at home should be introduced during a period of training at central hospitals, where the responsible district nurses were to stay longer than regular nurses and have access to a 'considerable amount of knowledge, both of the laws of health and of disease'.[89]

Rathbone opened the Liverpool Training School and Home for Nurses in Liverpool on 1 July 1862 and welcomed the new cohort of trainee district nurses that Nightingale had called for. Once students had completed their training, they were sent into the homes of the poor in one of eighteen districts in the city, each under the charge of a Lady Superintendent from a wealthy family. This organisational model proved successful in Liverpool and was soon expanded to other industrial cities, including Manchester (1865), Derby (1865), Leicester (1867), and to the East End of London (1868). The establishment of the Metropolitan and National Nursing Association for Providing Trained Nurses to the

Sick Poor (1875), headed by Florence Lees, marked the scheme establishing a national footing. The Queen's Nursing Institute (1890, QNI) then consolidated this structure ahead of state-supported district nursing in the twentieth century (Fig. 4.1).

Because of the ongoing presence of Lady Bountiful-style charity at the parish level, rural communities had less initial need for this urban style

Fig. 4.1 Cover of an early twentieth-century booklet on the Liverpool Queen Victoria District Nursing Association, one of the many district nursing organisations influenced by Nightingale's work with William Rathbone to provide care for the poor at home. *Queen's Nurses* (Liverpool: Liverpool Queen Victoria District Nursing Association, c. 1905), frontispiece, Wellcome Library, CMAC SA/QNI/X.38/1

of district nursing.[90] Elizabeth Malleson's Rural Nursing Association did seek to extend a similar scheme to the countryside, but its 1889 appeal to 'ladies who bury themselves in out-of-the-world villages' was in practice a reversion back to the language and image of the older, voluntary model.[91] In 1878 Nightingale expressed concern that by relying too heavily on these older practices, district nursing might 'degenerate into relief and easy visits, gossiping among the poor instead of gossiping among the rich'.[92]

Nightingale had always ranked the district nurse's task of putting 'the room into nursing order' equal to the actual treatment of illness. District nurses were to be models of cleanly habits, mandated to teach and encourage 'care and cleanliness' in all aspects of their patient's lives.[93] However, by the time that the philanthropist Angela Burdett-Coutts asked Nightingale to contribute to the 1893 World's Fair Congress on Hospitals, Dispensaries, and Nursing in Chicago, Nightingale had adopted the view that the educational aspects of home visiting should constitute a separate role.[94] In her paper for the event, Nightingale distinguished between 'nursing the sick'—a duty covered by the district nurse and assured of long-term support through the QNI—and 'health nursing', which she explained as 'the art of health, which every mother, girl, mistress, teacher, child's nurse, every woman ought practically to learn'. Returning to the concerns that had inspired *Notes on Nursing* many years earlier, Nightingale once again centred her efforts on shifting the domestic habits of 'the great mistress of family life'.[95]

In the 1890s Nightingale conveyed a deep scepticism in the belief that 'everything can be taught by book and lecture', drily noting that the 'tons of printed knowledge on the subject of hygiene and sanitation' had, in her view, done little to improve the population's health.[96] To address this gap between knowledge and application, Nightingale conjured a new form of woman visitor into being. She was aided in what became her 'Health at Home Mission', one of her last public campaigns, by Frederick Verney, the son of her brother-in-law, who as a county councillor had become Chair of the Buckinghamshire Technical Education Committee.[97] Together, Nightingale and Verney proposed a scheme that involved the Medical Officer for Buckinghamshire, Dr De'Ath, in the recruitment of what became known as health missioners to befriend and then educate rural inhabitants about matters of domestic sanitation. Like the Liverpool district nurses, these roaming educators were to receive structured training from well-educated women. As such, they embodied a departure from the uneven, haphazard, and unregulated sanitary education that rural populations had been subjected to in the past—a mere 'sprinkle' over the 'community', as Nightingale put it.[98] Although their teachings on hygiene and domestic planning were comparable to those set out in *Notes on Nursing*, health missioners were asked to adopt new pedagogical methods and encouraged to teach in a 'lively and dramatic manner', using the 'plainest, simplest, commonest words' and making 'the cottage itself ... be the object lesson'.[99] Their job title fused the religious zeal with which sanitary education had been associated since the middle of the century with the discourse of the imperial, so-called civilising mission that was in the ascendancy in these later decades.[100] Indeed, Nightingale proposed a very similar scheme of local educators, also named health missioners, to address village sanitation in India (Chapter 7).

Nightingale's health missioners scheme was a response to the educational, cultural, and social variations that threatened the effectiveness of public health campaigns across Britain. The primary challenge concerned the division between urban centres, where sanitary and medical expertise tended to cluster, and the rural hinterland, where knowledge about sanitation remained low and, in Nightingale's view, 'words ... go in at one ear and out at the other'. By drawing upon the supposedly unique capacities of 'women of the highest cultivation and of the deepest sympathy', health missioners were urged to follow 'the divine in their hearts to be "at home" in the cottage mother's homes'.[101] The description marked a curious re-emergence of the traditions of upper-class visiting within an increasingly systemised network of national health. And while Nightingale's health missioners remained as trials in Buckinghamshire, they closely resembled the model of health visiting then flourishing in Manchester and Birmingham, which became formalised and then expanded through the work of the Women Sanitary Inspectors' Association from 1896.[102]

Throughout her longstanding promotion of health at home, Nightingale had been a visitor to, writer for, and educator of the poor. In her early life the customary role of Lady Bountiful had provided a ready means of entering the lives of the less fortunate and of observing the conditions in which they lived. This role implicated the lady visitor in relationships of social deference and hierarchy whose motives Nightingale soon began to criticise, yet Lady Bountiful's presence subtly endured in her varied schemes to improve the sanitary conditions of homes. Nightingale's final campaign for health missioners recalled these traditions most directly, especially in terms of the faith placed in the virtues of well-educated women to benefit and improve the parishes with which they were most familiar. Proposing this scheme in the 1890s suggested a nostalgic wish that old social hierarchies and duties could meet the unprecedented challenges of the modern day—a sentiment shared by Burdett-Coutts, who insisted that '[t]he same kindly feelings that work to such noble effect in the Englishwoman of to-day animated the Englishwoman of yesterday', urging for 'the continuation and development [of their influence], under altered and more effective conditions'.[103] Meanwhile, Nightingale's vision of the district nurse treating and educating the poor in their homes had developed within a more organised infrastructure. The district nurse's rise was underwritten by the state's more active engagement with the intimacies of citizens' lives, replacing parochial traditions of care derived from feudal hierarchies with professional, institutionalised relationships.

Nightingale's varied effort to bring health into the home was not a wholly positive march to progress. If Lady Bountiful could be accused of exercising control as well as care through her visits, the measurements and observations undertaken by the more professionalised messengers that Nightingale later endorsed may have similarly felt, to those being educated and inspected, like intrusions into their homes. When Nightingale instead journeyed into the home through writing, the tensions that had existed between the visitor and the visited now came between the writer and the reader. Her remarkably popular *Notes on Nursing* was a laudable attempt to transform the living conditions of the population at large in the age of self-help, yet its advice could seem coercive, subtly imposing certain standards and behaviours on readers in their homes. Moreover, the insistence in this 'little book' that it was women who should lead the battle against disease in their homes was as limiting as it was empowering. On the one hand, *Notes on Nursing* promised women an exciting sphere of participation and social action, while on the other it reinforced a gendered association that tended to restrict their lives to the domestic space.

Notwithstanding these limitations, Nightingale's domestic mission made a significant and long-lasting contribution to what is today known as public health. While sanitary reformers had previously focused on changes to the law and local government, Nightingale's achievement was to bring the crusade against filth into the everyday home. It was a work of translation. The aim, at least, was to present the latest theories of disease and prevention in ways that addressed and enthused the everyday lives of the British population. Whether finding inventive methods of writing or training new kinds of professionals to meet this task, Nightingale's lifetime commitment to health at home inspired some of her most original and influential work.

Notes

1. Karen Harvey, *The Little Republic: Masculinity and Domestic Authority in Eighteenth-Century Britain* (Oxford: Oxford University Press, 2012), 3.
2. Judith Flanders, *The Victorian House: Domestic Life From Childbirth to Deathbed* (London: HarperCollins, 2003), 286–301; Amy Partridge, 'Public Health for the People, the Use of Exhibition and Performance to Stage the "Sanitary Idea" in Victorian Britain', PhD diss., Northwestern University, 2005.
3. Tom Crook, *Governing Systems: Modernity and the Making of Public Health in England, 1830–1910* (Oakland: University of California Press, 2016), 16.
4. Jessica Gerard, 'Lady Bountiful: Women of the Landed Classes and Rural Philanthropy', *Victorian Studies* 30, no. 2 (1987): 183–210, 182; see also Pamela Horn, *Ladies of the Manor* (Stroud: Amberley, 2014). On the effects of such visits on a working-class sense of privacy see Martin Hewitt, 'District Visiting and the Constitution of Domestic Space in the Mid-Nineteenth Century', in *Domestic Space: Reading the Nineteenth-Century Interior*, ed. Ingra Bryden and Janet Floyd (Manchester: Manchester University Press, 1999), 121–141.
5. Carolyn Steedman, 'What a Rag Rug Means', in *Domestic Space*, ed. Bryden and Floyd, 21.
6. Emily Cuming, *Housing, Class and Gender in Modern British Writing, 1880–2012* (Cambridge: Cambridge University Press, 2016), 25–35.
7. Henry Mayhew, 'Home Is Home, Be It Never So Homely', in *Meliora: Or, Better Times to Come*, ed. Viscount Ingestre (London: John W. Parker, 1853), 263–264.
8. Nightingale, 'Lebenslauf', 24 July 1851, in *The Collected Works of Florence Nightingale, Volume 1: Florence Nightingale: An Introduction to Her Life and Family*, ed. Lynn McDonald (Waterloo, ON: Wilfrid Laurier University Press, 2001) [hereafter *CW* 1], 92.
9. Nightingale to Frederick Verney, 17 October 1891, in *The Collected Works of Florence Nightingale, Volume 6: Florence Nightingale on Public Health Care*, ed. Lynn McDonald (Waterloo, ON: Wilfrid Laurier University Press, 2004) [hereafter *CW* 6], 587.

10. '"Rural Hygiene", a Paper by Florence Nightingale Read by Frederick Verney', in *Official Report of the Central Conference of Women Workers* (1894), 46–60, *CW* 6, 619–620.
11. K. D. Reynolds, *Aristocratic Women and Political Society in Victorian Britain* (Oxford: Clarendon Press, 1998), 103.
12. Nightingale to Harry Verney, 2 April [1861 or 1862], in *The Collected Works of Florence Nightingale, Volume 5: Florence Nightingale on Society and Politics, Philosophy, Science, Education and Literature*, ed. Lynn McDonald (Waterloo, ON: Wilfrid Laurier University Press, 2003) [hereafter *CW* 5], 171.
13. From Nightingale's lost childhood memoirs, 1830, quoted in I. B. O'Malley, *Florence Nightingale, 1820–1856: A Study of Her Life Down to the End of the Crimean War* (London: Thornton Butterworth, 1931), 33.
14. Frank Prochaska, 'Women in English Philanthropy, 1790–1830', *International Review of Social History* 19, no. 3 (1974): 426–445.
15. Hannah More, *Coelebs in Search of a Wife* (London: Cadell and Davies, 1808), 356.
16. Christopher Benson, *The District Visitor's Manual* (London: John W. Parker, 1840), vi.
17. See Dorice Williams Elliott, *The Angel Out of the House: Philanthropy and Gender in Nineteenth-Century England* (Charlottesville: University of Virginia Press, 2002), 60–61; Martin Gorsky, *Patterns of Philanthropy: Charity and Society in Nineteenth-Century Bristol* (Woodbridge: Boydell Press, 1999).
18. Note in response to a letter by Elizabeth Eastlake, April 1868, *CW* 5, 259.
19. Maria Coape to Fanny Nightingale, 18 March 1834, Claydon N15/28.
20. Julia Smith to Fanny Nightingale, undated [1834], Claydon N15/31.
21. Note by William Nightingale, 'Lea Hurst, Thursday Evening', undated, Claydon N73.
22. Maria Coape to Fanny Nightingale, 18 March 1834, Claydon N15/28.
23. Anne Summers, 'A Home from Home: Women's Philanthropic Work in the Nineteenth Century', in *Fit Work for Women*, ed. Sandra Burman (London: Croom Helm, 1979), 33–63, 48.
24. Note by Nightingale, *CW* 1, 56.
25. Parthenope Nightingale to Mary Shore, July 1836, quoted in Mark Bostridge, *Florence Nightingale: The Woman and Her Legend* (London: Penguin, 2009), 50.
26. Bostridge, *Florence Nightingale*, 50–51.
27. Nightingale, 'Lebenslauf', *CW* 1, 91.
28. Nightingale to her mother, undated [1851], in *The Collected Works of Florence Nightingale, Volume 7: Florence Nightingale's European Travels*,

ed. Lynn McDonald (Waterloo, ON: Wilfrid Laurier University Press, 2004) [hereafter *CW* 7], 684.
29. Nightingale to her sister, 12 February 1837, *CW* 1, 286.
30. Anthony Wohl, *Endangered Lives* (London: Dent, 1983), 128.
31. Elizabeth Gaskell, *The Letters of Mrs Gaskell*, ed. J. A. V. Chapple and Arthur Pollard (Manchester: Manchester University Press, 1966), 306.
32. Nightingale's diary, 16 July 1846, quoted in O'Malley, *Nightingale*, 119–120.
33. Nightingale to Fred Verney, 23 June 1879, *CW* 1, 743.
34. Nightingale to Fred Verney, 25 October 1881, LMA (FNM) f45v; see also Joyce Schroeder MacQueen, 'Florence Nightingale's Nursing Practice', *Nursing History Review* 15 (2007): 29–49.
35. See Priscilla Wald, *Contagious: Cultures, Carriers, and the Outbreak Narrative* (Durham, NC: Duke University Press, 2008).
36. Beatrice Smith to Edwin Chadwick, 18 September [1858], *CW* 1, 523–533.
37. See Lynn McDonald, *Florence Nightingale at First Hand* (London, Continuum, 2010), 101–3. As McDonald has shown, Nightingale later changed her position on germ theory, accepting it once Robert Koch had isolated the pathogens for tuberculosis and cholera in 1882–1883.
38. Nightingale, *Notes on Nursing, and Notes on Nursing for the Labouring Classes: Commemorative Edition with Historical Commentary*, ed. Victor Skretkowicz (London: Harrison, 1860 and 1861; New York: Springer, 2010), 108. Citations are to the Springer edition. This chapter refers to two distinct versions of Nightingale's book. First, *Notes on Nursing: What It Is, and What It Is Not* (July 1860, often incorrectly given as 1859), or the Library Standard edition ('lib. ed.'), an expanded version of the earliest version of the text published in January 1860. Second, *Notes on Nursing for the Labouring Classes* (April 1861), or the Labouring Classes edition ('lab. ed.'), substantially revised to appeal to a popular audience.
39. Charles E. Rosenberg, *Explaining Epidemics and Other Studies in the History of Medicine* (Cambridge: Cambridge University Press, 1992), 92.
40. Nancy Aycock Metz, 'Discovering a World of Suffering: Fiction and the Rhetoric of Sanitary Reform', *Nineteenth-Century Contexts* 15, no. 1 (1991): 65–81, 65.
41. For a traditional account see R. A. Lewis, *Edwin Chadwick and the Public Health Movement 1832–1854* (London: Longman, 1952) and for revisionist accounts see Mary Poovey, *Making a Social Body: British Cultural Formation, 1830–1864* (Chicago: University of Chicago Press, 1995), 115–131; Christopher Hamlin, *Public Health and Social Justice*

in the Age of Chadwick: Britain, 1800–1854 (Cambridge: Cambridge University Press, 1998).

42. Elaine Freedgood, *Victorian Writing About Risk: Imagining a Safe England in a Dangerous World* (Cambridge: Cambridge University Press, 2009), 50.
43. Edwin Chadwick to Nightingale, 8 December 1858, BL Add MS 45770 ff89–90.
44. It was left to Harriet Martineau's *England and Her Soldiers* (1859) to address a popular audience with these findings. See Iris Veysey, 'A Statistical Campaign: Florence Nightingale and Harriet Martineau's *England and Her Soldiers*', *Science Museum Group Journal* 5 (2016), https://doi.org/10.15180/160504.
45. Chadwick to Nightingale, 8 December 1858, BL Add MS 45770 ff89–90.
46. Bessie Rayner Parkes, 'The Ladies' Sanitary Association', *The English Woman's Journal* 3, no. 14 (1859): 81–82.
47. See Poovey, *Social Body*, 115–131.
48. Rosenberg, *Epidemics*, 99.
49. Journal entry quoted in O'Malley, *Nightingale*, 127.
50. Nightingale, *Notes on Nursing*, lib. ed., 49.
51. 'Miss Nightingale's *Notes on Nursing*', *Quarterly Review* 107, no. 214 (1860): 404.
52. Patricia Branca, *Silent Sisterhood: Middle-Class Women in the Victorian Home* (London: Croom Helm, 1975).
53. Nightingale, *Notes on Nursing*, lib. ed., 50.
54. Samuel Smiles, *Self-Help* (London: John Murray, 1859), 1.
55. Samuel Smiles, *Character* (London: John Murray, 1871), 40.
56. John Sutherland to Nightingale, 10 February 1859, BL Add Mss 45751 ff125b–126.
57. Nightingale, *Notes on Nursing*, lib. ed., 64, 70.
58. Nightingale to Harriet Martineau, 29 July 1860, in *The Collected Works of Florence Nightingale, Volume 8: Florence Nightingale on Women, Medicine, Midwifery and Prostitution*, ed. Lynn McDonald (Waterloo, ON: Wilfrid Laurier University Press, 2005) [hereafter *CW* 8], 613.
59. Nightingale, *Notes on Nursing*, lib. ed., 95; Sutherland to Nightingale, 12 February 1859, BL Add Mss 45757 ff245.
60. Nightingale, *Notes on Nursing*, lib. ed., 85, 68, 71–72.
61. On Nightingale's 'rhetorical and narrative strategies' see Louise Penner, *Victorian Medicine and Social Reform: Florence Nightingale Among the Novelists* (New York: Palgrave Macmillan, 2010), 76.
62. Nightingale, *Notes on Nursing*, lib. ed., 85, 71.
63. Skretkowicz, 'Introduction', in Nightingale, *Notes on Nursing*, 14.
64. 'Nightingale's *Notes*', *Quarterly Review* 107, no. 214 (1860): 393.

65. 'Miss Nightingale's Notes', *The Times*, 24 January 1860, 7.
66. Chadwick to Nightingale, 14 January 1860, *CW* 6, 18.
67. Review quoted in *The Saturday Review*, 28 April 1860.
68. Ibid.
69. Nightingale, *Notes on Nursing*, lab. ed., 349, 328, 315; lib. ed., 97.
70. Nightingale to her mother, 21 April 1861, *CW* 1, 149.
71. Skretkowicz, 'Introduction', 24.
72. William Farr to Nightingale, 29 April 1861, *CW* 6, 19.
73. Michelle Allen, 'From Cesspool to Sewer: Sanitary Reform and the Rhetoric of Resistance, 1848–1880', *Victorian Literature and Culture* 30, no. 2 (2002): 383–402, 396.
74. Nightingale to Amy Hawthorn, 6 November 1891, *CW* 8, 401.
75. Parkes, 'Ladies' Sanitary Association', 84, 74, 85.
76. S. R. Powers, 'The Diffusion of Sanitary Knowledge', *Transactions of the National Association for the Promotion of Social Science* (1860): 715–716; 'Sanitary Sermons', *Punch*, 13 January 1872, 15. See also Alison Bashford, *Purity and Pollution* (London: Macmillan, 1998), 12–14.
77. Wohl, *Endangered*, 66.
78. Benjamin Ward Richardson, 'Woman as a Sanitary Reformer', *Transactions of the Sanitary Institute* (London: Sanitary Institute, 1880), 188.
79. Parkes, 'Ladies' Sanitary Association', 82.
80. Ibid., 84; for context see Eileen Janes Yeo, 'Social Motherhood and the Sexual Communion of Labour in British Social Science, 1850–1950', *Women's History Review* 1, no. 1 (1992): 63–87.
81. *The Lancet*, 19 April 1862.
82. Octavia Hill, *Our Common Land* (London: Macmillan, 1877), 59–60.
83. Gaskell, *Letters*, 319–320.
84. Edward H. Sieveking, 'On Dispensaries and Allied Institutions', in *Lectures to Ladies*, ed. F. D. Maurice (Cambridge: Macmillan, 1855), 91–116, 116, 94.
85. 'Ladies' Association for the Promotion of Sanitary Knowledge', *The Philanthropist*, 1 August 1859, 434.
86. Classic histories of district nursing include Mary D. Stocks, *A Hundred Years of District Nursing* (London: Allen & Unwin, 1960); Gwen Hardy, *William Rathbone and the Early History of District Nursing* (Ormskirk: Hesketh, 1981).
87. Anne Summers, 'The Mysterious Demise of Sarah Gamp: The Domiciliary Nurse and Her Detractors, c. 1830–1860', *Victorian Studies* 32, no. 3 (1989): 365–386.
88. Nightingale to William Rathbone, 13 August 1860, in *The Collected Works of Florence Nightingale, Volume 13: Florence Nightingale: Extending Nursing*, ed. Lynn McDonald (Waterloo, ON: Wilfrid Laurier University Press, 2009) [hereafter *CW* 13], 255.

89. Nightingale, 'Introduction', in William Rathbone, *Organization of Nursing: An Account of the Liverpool Nurses' Training School* (Liverpool: Holden, 1865), 9–16, *CW* 6, 261, 255.
90. Carrie Howse, 'From Lady Bountiful to Lady Administrator: Women and the Administration of Rural District Nursing in England, 1880–1925', *Women's History Review* 15, no. 3 (2006): 423–441, 427.
91. Elizabeth Malleson writing in the *Nursing Record*, 17 October 1889, 232.
92. Nightingale to Captain Fortescue, 19 February 1878, *CW* 13, 759.
93. Nightingale, 'Training Nurses for the Sick Poor', *The Times*, 14 April 1876, *CW* 13, 752, 753.
94. Louise C. Selanders and Patrick Crane, 'Florence Nightingale in Absentia: Nursing and the 1893 Columbian Exposition', *Journal of Holistic Nursing* 28, no. 4 (2010): 305–312.
95. Nightingale, 'Sick-Nursing and Health-Nursing' (1893), in *CW* 6, 203–219, 206.
96. Ibid., 210.
97. Nightingale to Frederick Verney, 21 November 1892, *CW* 13, 885.
98. Nightingale, 'Sick-Nursing and Health-Nursing', *CW* 6, 212.
99. Nightingale to Guthrie Wright, 20 May 1892, *CW* 13, 866; Nightingale to Rose Adams, 17 February 1892, *CW* 13, 881.
100. Nightingale's proposal had similarities with the earlier 'Sanitary Home Missionaries' employed by the Ladies' Sanitary Association from the late 1850s. See Powers, 'Diffusion of Sanitary Knowledge', 714.
101. Nightingale, 'Rural Hygiene', *CW* 6, 620, 621.
102. Celia Davies, 'The Health Visitor as Mother's Friend: A Woman's Place in Public Health, 1900–14', *Social History of Medicine* 1, no. 1 (1988): 39–59.
103. Angela Burdett-Coutts, *Woman's Mission: A Series of Congress Papers on the Philanthropic Work of Women* (London: Sampson Low, 1893), xiv.

CHAPTER 5

Homely Institutions

'She is, I think, too much for institutions', Elizabeth Gaskell wrote of Florence Nightingale in 1854.[1] In Gaskell's eyes, Nightingale's emphasis on institutions as sites for achieving reform went against orthodox Victorian ideas of the unparalleled value of hearth and home. Yet in another sense, the comment further exemplifies how Nightingale was thoroughly a woman of her time, for despite its reverence for domesticity, the nineteenth century was an age of institution building. As the population, economy, and public sector expanded, so did the number of schools, hospitals, lunatic asylums, workhouses, prisons, universities, religious missions, and lodging houses. Rather than just inert structures, institutions became a way of shaping how people—especially poorer people—thought, behaved, and interacted.[2] The new buildings formed a web of authority, order, and power that contributed to configuring society; public architecture was intimately connected with the thinking and morality of the nation.[3] An articulation of this idea can be found in Nightingale's writings: the 'public hospitals of any country', she wrote, indicated the 'standard of the knowledge' of its government and the level 'of civilisation amongst a people'.[4] For Nightingale, as for many of her contemporaries, good standards of public buildings and housing were necessary to ensure high standards of moral conduct and of public health.

Over the course of her working life, Nightingale became involved with hundreds of institutions: designing and redesigning hospitals, reforming workhouses, creating nurse training schools and nurses' residences. This

P. Crawford et al., *Florence Nightingale at Home*,
https://doi.org/10.1007/978-3-030-46534-6_5

chapter focuses on three key institutions that Nightingale, respectively, attended, managed, and created: the Deaconess Institution at Kaiserswerth, Germany; the Establishment for Gentlewomen During Illness in Harley Street, London; and the Nightingale School at St Thomas' Hospital. These institutions reflected contemporary developments in nursing and, in one way or another, were influenced by the design, moral associations, and organisational structures of home.

Homely Institutions and the Quest for 'Respectable' Nurses

As earlier chapters have discussed, ideals of home and domesticity reached new heights of cultural influence in the mid-Victorian period. Middle-class Victorians romanticised and idealised the ordered, gendered home environment, projecting it outwards as a model for broader social morality. This thinking, inevitably, also infused many new public institutions, which took on the language and values of the Victorian household. Jane Hamlett's book *At Home in the Institution* (2015) shows how 'domestic material culture and practices' were used to 'transform institutional environments' such as mental asylums, schools, and lodging houses. Institutional authorities sought to recreate elements of the middle-class home by, for example, placing parlour ornaments in asylum wards, or by including a drawing room or a reading room with 'cosy corners' for patients.[5]

Questions were raised about the extent to which middle-class women should be involved in the new institutions. On the one hand, British society was predicated on ideals of gendered segregation in which respectable women were expected to concentrate on producing homes infused with moral virtue. This implied that middle- and upper-class women should be kept away from institutions in the public realm like hospitals and workhouses. On the other hand, visiting institutions was a logical extension of charitable parish visiting by upper-class women, since philanthropic ladies also assumed a social responsibility for reforming the morals, manners, and behaviour of the poor, attenuating social ills through the power of their feminine presence (Chapter 4). These contrasting discourses were further complicated by women like Nightingale insisting that their 'passion for meaningful work' should find an outlet, broader than that expected through voluntary social duty.[6]

A great challenge for middle-class women in the nineteenth century was to find ways to work in the public realm without losing respectability. Ultimately, as Ellen Jordan has shown, this required a discursive shift in how the proper role and domain of women were understood. Reformers of the 1840s and 1850s used religious arguments to shift middle-class common sense but had to be careful not to transgress conventional sensibilities too far.[7] Reforming employers therefore sought to make the visible aspects of working life such as living arrangements, dress, and training appear respectable according to the gendered expectations of the time, thereby curating a space for women to work without risking social condemnation. In the case of nursing, recent research has suggested that rather than the Nightingale-era reforms resulting in the replacement of working-class nurses with middle-class ones (as was often previously assumed) it was rather the social *status* of nursing that was transformed, from a menial service job into a respectable profession.[8] This shift in status was brought about by either working within or adapting contemporary ideas about femininity. As Eva Gamarnikow has written, 'nursing reformers employed ideologies of femininity in an enabling manner', constructing nursing as 'femininity in action' in order to justify creating new career paths for women.[9]

While the three institutions discussed below belong to distinct periods of Nightingale's life, they also form part of a wider story about the evolution of nursing from a domestic service occupation into a health-care profession.[10] As new scientific understandings displaced older ideas about restoring patient balance, and new medical technologies (such as the stethoscope) began to replace older subjective diagnostics with a form of empirical precision, so new treatments like vaccinations and the administration of pain-relieving opiates demanded more intensive and skilled nursing care. Similarly, new techniques of surgery (supported by the advent of anaesthetics and antiseptics) placed new emphasis on cleanliness in operating theatres and during post-operative recovery. These changes not only revolutionised medicine but created a demand for supportive staff that was not matched to the skills of eighteenth-century nurses. The hospital nurses of 1800 were recruited from the more downtrodden ranks of the servant class, since hospital work was considered less desirable than domestic service. Like housemaids, nurses generally slept on site, often in the wards or corridors of hospitals. Their hours—typically from 6 a.m. to 10 p.m., seven days a week—were significantly longer than those of factory workers. They had to do their personal washing on the wards, and had few pleasures in life beyond alcohol, which was readily available. By the 1820s, London doctors and hospital administrators were complaining that their nurses were not up to the new requirements. The emphasis on cleanliness and the new care tasks required nurses 'with a good character, and with something of an education', wrote one hospital governor in 1824. Westminster Hospital's governors in 1838 similarly sought to begin to hire 'a superior class of woman' as ward sisters.[11]

Despite this clearly perceived need, hospital managers struggled to find nurses with the requisite qualities. The public reputation of nursing was low, as famously exemplified in Charles Dickens's caricatures of Sarah Gamp and Betsy Prig, or as shown in Nightingale's own 1850 statement that most contemporary hospitals were places of 'immorality and impropriety', where 'women of bad character are admitted as nurses, to become worse by their contact with the male patients and the young surgeons'.[12] Hospital nurses, typically aged in their later thirties or forties, were seen by middle-class moralisers as impure, uncouth, and unchaste—the antithesis of the domestic ideal. Middle-class Victorians thought nurses needed more education, but that above all they needed to become more reliable, punctual, sober, chaste, honest, and industrious.

To achieve respectability for nurses, it was imperative that they no longer live on the wards. This was a question of self-respect as well as social reputation: nurses were unlikely to feel a strong commitment to the work while sleeping in wooden cages on a hospital landing, as Nightingale observed at St Bartholomew's.[13] The first organisations to provide more attractive living arrangements for women undertaking nursing work were religious sisterhoods. Various Catholic sisterhoods sprung up after the 1829 Emancipation Act, notably the Bermondsey convent of the Irish Sisters of Mercy led by Mary Clare Moore. From the 1840s, Anglican sisterhoods encouraging nursing began to appear, including the Park Village Sisterhood and the Devonport Sisters of Mercy, both of which sent nurses to Scutari with Nightingale in 1854, Elizabeth Fry's Institution of Nursing Sisters, and the All Saints Sisters of the Poor, which took over University College Hospital's nursing service in 1862. Most significant of all was St John's House, founded in 1848, which stood out for its relatively loose religious commitment and resemblance to a university college for women. Under the dynamic Mary Jones, superintendent from 1853, St John's became a real leader of nursing reform and undertook the nursing at King's College Hospital from 1856. Jones's battle to wrest disciplinary authority over the nurses from the male house 'master' (a chaplain) anticipated similar debates affecting Nightingale's own institutions.[14]

These sisterhoods addressed the problem of nurses' miserable living conditions by setting up comfortable convents or community homes in salubrious districts, a safe distance away from the hospitals. Their strict behavioural codes, daily routines, and supervision by chaplains and lady superintendents guaranteed their outward social respectability. Beneath this outward conservatism, they created something radical: communities of single women, dedicated to social work, largely independent from male control. Protected by the Church and a uniform, women could act in ways and go to places that otherwise would have been forbidden to them.[15] Furthermore, sisterhood houses were places in which women could find privacy and time for self-development—the very things that Nightingale found so lacking in her family homes. As such, they provoked anxieties for some contemporary observers who feared that women who entered sisterhoods were running away from their home duties.[16] Partly as a defence against such criticisms, sisterhoods cultivated a family atmosphere. Religious homes were typically governed by a 'Mother' who watched benevolently over her team of Sisters, with a chaplain or male governor acting as a distant father figure.

Sisterhoods thus pioneered the drive towards reformed, respectable, trained nursing, enlarging the possibilities for nineteenth-century women's work in the process. Nightingale's earliest ideas about reforming nursing were bound up with this model. After her parents refused her permission to learn practical nursing in Salisbury Hospital in 1845, she explained to her cousin that her longer-term idea had been 'a fine plan… of taking a small house in West Wellow… I thought something like a Protestant Sisterhood, without vows, for women of educated feelings, might be established'.[17] When she left for Scutari in October 1854, twenty-four of the thirty-eight nurses in her party were from religious orders, indicating her continued belief in their merits.[18] She did not spend significant time in any of the Anglican orders, though Mary Jones, whom she met in October 1854 in the flurry of preparation to depart for the Crimean War, became an important friend and colleague.[19] While the sisterhoods in Britain had provided an important precedent, Nightingale's greatest source of influence lay, instead, in Germany.

The 'Motherhouse': Kaiserswerth

Nightingale's relationship with the Deaconess Institution at Kaiserswerth, in the Rhineland, belongs to her period of intense frustration at the restrictions of country house life (Chapter 3). She became aware of Kaiserswerth in 1846, when one of its annual reports was brought to her home by a family friend, the Prussian ambassador Christian von Bunsen.[20] By that date, Kaiserswerth was becoming a source of inspiration, and something of a pilgrimage site, for Protestant women from around Europe who hoped to perform socially useful work without sacrificing respectability. The ambassador had set up a Protestant infirmary in Rome and a German hospital in London (which Nightingale visited) using Kaiserswerth-trained nurses.[21] In 1848, Nightingale noted with interest that a Swiss woman 'of rank and education' had gone from Berne 'to Kaiserswerth to learn', creating on her return a small hospital which she used to train other women in nursing.[22] This was exactly the kind of project that Nightingale had in mind for herself at that time.

The Deaconess Institution had been set up in the 1830s by an evangelical pastor, Theodore Fliedner, and his wife Friederike. It was partly British in inspiration: Fliedner had visited England in 1824, meeting the Quaker social reformer Elizabeth Fry at Newgate prison.[23] In 1833, Fliedner set up a rehabilitation home for female convicts in his Kaiserswerth parish. This subsequently expanded into a multifaceted institution, where women could train to become evangelical deaconesses skilled in one or more aspects of health care or social work. With the support of Friederike and later his second wife Caroline, Fliedner opened a small hospital in 1836, and further added a school, an orphanage, and a teacher training college. All this sat among forty acres of land, on which the deaconesses grew vegetables and kept cows and horses. Therefore, although the institution became famous as a hospital—one named after Nightingale now stands on the site—Kaiserswerth was always a much broader organisation, devoted not only to health care but also to social work of various kinds. From the 1850s, the institution, by that point partly funded by the Prussian state, began to expand, opening further sites across the German Confederation and in the Middle East and the Balkans.[24]

The crucial achievement of Kaiserswerth was that it allowed young, unmarried, Protestant women to work in health care and social work without losing caste, thereby creating a model for a life outside marriage and family duty. The Fliedners did this by creating a community in which they, as directors, took on a form of parental responsibility towards the deaconesses, imposing a strict set of house rules that protected their charges' legal security and social respectability. They introduced a uniform—floor-length, dark-blue dresses made from expensive cloth, topped off by lace bonnets—that gave nurses the appearance of honourable married women. They hired five male nurses to work in the men's wards, so that no female nurse needed to interact intimately with male patients. Deaconesses were required to subject themselves to a thorough self-examination that stressed duty, humility, and self-denial, while adopting a subordinate position towards the doctors and pastor. Women who successfully internalised these attitudes, it was hoped, would gradually bring the patients, convicts, prostitutes, and orphans among whom they worked into the realm of respectable society and Christian piety (Fig. 5.1).[25]

Fig. 5.1 Painting of Fliednerstrasse, Kaiserswerth, Germany at the time that Nightingale first visited the Deaconesses' Institute. Johann Baptist Sonderland, 1850, Fliedner Kulturstiftung Kaiserswerth

Nightingale visited Kaiserswerth twice: in August 1850 on her way home from travelling in Egypt and Greece, staying for two weeks, and for a longer, three-month spell in 1851. At the end of her first stay, Fliedner asked her, as part of her service to the institution, to write an English-language history of it for popular distribution. The resulting twenty-page text, quickly completed in August 1850 and published in 1851, remains an important source on the history of Kaiserswerth. It included a significant introductory section on the restrictions on Englishwomen's ability to undertake meaningful work that puts forward inspirational examples of past and present communities of deaconesses to overcome these. Nightingale's diary and notes from her 1851 visit also contain important material on her experiences there.[26]

Even before visiting Kaiserswerth, Nightingale had identified it as 'a home, to young women willing to come'.[27] Nightingale likely meant 'home' here in the spiritual sense, associated with her desire to work, that we saw in her reading of Thomas Carlyle (Chapter 2). Kaiserswerth was a home, not because it provided the comforts of Victorian domesticity, but because it offered young women purpose and training to prepare them for specific kinds of work. Indeed, Nightingale's friend Selina Bracebridge described Kaiserswerth as 'a rough place', a 'coarse practical affair', with a 'want of "comforts"' when she visited with Nightingale in 1850.[28] During her 1850 stay, Nightingale struggled to so much as obtain a tub of cold water to wash in.[29] She admitted even to her parents, who objected to her going there in the first place, that the hospital was 'by no means a pattern of cleanliness'.[30] Life at Kaiserswerth was simple, humble, austere, but despite—or perhaps because of—the material discomforts, there was a strong sense of community among the deaconesses and trainees, who all lived and worked together.[31] At Kaiserswerth, wrote Nightingale, 'there are, for *all*, the same privations, the same self-denial, the same object: one spirit, one love, one Lord'.[32]

Furthermore, the trainees were encouraged to think of themselves as part of a family, and to call the superintendent, Caroline Fliedner, 'Mother'. 'It is beautiful to see the attachment which the deaconesses of Kaiserswerth feel to their "mother house"', Nightingale wrote in 1850.[33] The children in the school and orphanage were made to feel valued—every child's birthday was celebrated 'with flowers, telling stories, singing'.[34] During her 1851 visit, Nightingale undertook various community activities unconnected to the hospital, including taking schoolchildren on walks, giving English lessons, and undertaking fund-raising ('begging') excursions.[35] She contrasted the uplifting sense of vocation and community among Kaiserswerth deaconesses with the 'unutterable dullness' of British hospitals, where nurses performed their tasks 'for hire and not for love'.[36] None of Kaiserswerth's women's wards contained more than four beds, ensuring patients' privacy and dignity were respected—again in notable contrast to British hospitals, where neither nurses nor patients enjoyed much privacy.[37] Nightingale praised 'the delicacy, the cheerfulness, the grace of Christian kindness, the moral atmosphere, in short' of Kaiserswerth, describing it as 'humanising, refining, propriety-teaching'; 'one of God's schools'.[38] But unlike other Godly schools, such as the Catholic sisterhoods on which it was partly based, its participants were not signing away the rest of their lives, nor did they live in convent-style cloisters. Deaconesses typically made a commitment to train for three years and serve for five but could ask for release at any time and for any reason.[39]

Kaiserswerth can be understood in terms of what Eileen Yeo has called 'social motherhood'—the idea that women who chose not to follow the conventional path of home and family life nonetheless 'did not abandon the vital elements of motherhood and home but rather enlarged and altered these to fit the single woman'. Instead, they created 'a new image of a virgin mother engaged in self-sacrificing work with the poor and needy in the public world and introducing a home influence into it'.[40] This idea was fundamental to the self-conception of the Kaiserswerth deaconesses and, after 1854, became a vital component of how Nightingale's public image was constructed, as Chapter 6 will show. These ideas also came to underpin Nightingale's conceptions of the Nightingale School and Nightingale Home at St Thomas' in her later years.

A further important feature distinguishing Kaiserswerth from British voluntary hospitals was that instead of doctors being in ultimate charge of the infirmary, 'the clergyman [i.e. Fliedner] is master'.[41] Nightingale largely supported this system, agreeing that even though the doctors' orders should always be followed on medical matters, it was important that the nurses should instead be accountable to a separate authority. While Fliedner fitted this non-medical mould, she nonetheless considered him something of an authoritarian patriarch, and would have preferred a woman to be in charge.[42] On Fliedner's death in 1864 she argued that Caroline, as 'Mother' of the deaconesses and former director of nursing at Hamburg General Hospital, should take over rather than his son-in-law.[43] The importance of an independent women-led hierarchy for nursing was a principle that Nightingale was repeatedly to insist upon in later years.

During her 1851 visit, Nightingale spent only part of her time working in Kaiserswerth's infirmary—she was often assigned to other parts of the Institution, notably the apothecary, spending numerous days making up medicines. Even the 'ward' days involved various other duties, as described in her 1850 diary:

Wednesday 16 July

Women's ward
6.15 a.m. On duty, the other sisters at breakfast.
7–9 a.m. Bedmaking, sweeping, combing, dressing wounds. Rubbing in ointment, the cod liver oil taken as early as possible …
9–9.30 a.m. I held prayers to the patients collected, while the other sister was doing their rooms …
10. a.m. Doctor's visit.
11–12.30 a.m. Two sulphur baths for Frau Marcus and Frau Brose. …
1.30–2 p.m. Told them stories about Athens.
…
3–4 p.m. I with three women to church, the others in the Kirchenzimmer, the rest in bed.
5–6 p.m. Took Adelheit Schulz … a walk, first shopping in Kaiserswerth, then along the Rhine to the old ruin – very striking, the broad flowing river.
7.15–8 pm. Verbinden [bandages]
…

Saturday 2 August

Men's ward
4.30–6 a.m. Did the *Zionsverein* [Kaiserswerth offshoots being set up by Fliedner in Palestine] for Pastor.
6.00–6.15 a.m. Distributed breakfast.
…
8.30–10 a.m. Visits from doctor. Prayers as usual. The amputated man going on well – much with him: wet compresses every ten minutes. He was never left alone. Read with him.
…
8–9 p.m. English lesson [i.e. given by Nightingale].[44]

At Kaiserswerth, Nightingale learned practical nursing techniques such as compresses, dressing wounds, giving enemas, applying leeches, and cupping, as well as how to lay out the dead.[45] She does not seem to have attended academic-style lessons—the Fliedners offered a 'medical course' providing basic training in anatomy, but few lectures seem to have taken place during her visits.[46] As these extracts make clear, however, the Fliedners did not simply provide training, but immersed even short-term visitors in a wide range of activities and expected them to make varied contributions to community life as befitted their individual skills and aptitudes.

It was above all this immersive, busy experience of working with patients in a supportive and religiously motivated community that, Nightingale wrote, 'fills my life with interest and strengthens me body and mind'.[47] Community bonds among the deaconesses were developed through regular initiation and confirmation ceremonies and by encouraging graduates to maintain correspondence with, and return frequently to, the motherhouse. The Fliedners placed much emphasis on prayer and on religious and moral instruction, seeming to value this over intelligence and social refinement.

Relatively short though it was, Nightingale's Kaiserswerth experience was enough to reconfirm her sense of nursing vocation. She saw her experiences there as useful and energising, in sharp contrast to her prior life of, to borrow her own terms, idleness and daydreaming. 'This it is to live', she enthusiastically wrote in her diary, after spending two hours holding compresses to the temples of the 'amputated man', as he later lay dying of typhus.[48] Or as she wrote a couple of weeks later, 'it is astonishing how much good a stern reality does one'.[49] She left Kaiserswerth with clearer ideas about how nursing institutions could be run and what kind of people should be recruited. Experience of Kaiserswerth's communal living left her convinced that moral character and religious purpose mattered much more than social background. As she wrote to her father: 'I cannot tell you how much I thought of many of the sisters with whom I was in the closest contact, almost all out of the lower classes … I saw what power the[ir] having devoted all to God has in refining the intellect and giving grace to the character'.[50] Yet if class mattered less, Nightingale's 1850 pamphlet nevertheless argued that a crucial ingredient of Kaiserswerth's success was its absence of monetary inducements—'nothing is offered to the sisters, neither the prospect of saving money, nor reputation, nothing but the opportunity of working in the cause for which Christ worked … if this does not appear to be their ruling principle, they are dismissed'.[51] While Nightingale accepted that women without independent wealth should be paid to nurse, she always believed that the work itself should be the real reward.

In the context of a severely restricted world of professional opportunities for women, Kaiserswerth in 1850 stood out for its ability to act as a community, home, and motherhouse that sent deaconesses out into the world and welcomed them back in times of need. It presented an alternative to the family that nevertheless retained its ties of loyalty, and indeed, by practising social motherhood, it encouraged its pupils to direct motherly and familial feelings in the direction of the poor, ill, and needy. As Nightingale put it in *Suggestions for Thought*, it was the sort of institution that would 'not destroy the family, but make it larger'.[52]

Putting the House in Order: The Establishment for Gentlewomen During Illness

Kaiserswerth was a physical manifestation of Nightingale's belief that the right kind of religious ethos could lead to new, work-orientated communities for women. Her challenge for at least the next decade was to establish her own version of this new sort of working home. Her first attempt, the Establishment for Gentlewomen During Illness in Harley Street, London, was only a partial success, since Nightingale was unable to recruit either the patients or the nurses necessary to generate a Kaiserswerth-style home or community. Nonetheless, her experience there saw her wrestling in new ways with the idea of home, and how this should—or should not—be incorporated into healthcare institutions.

One of the things that Nightingale liked about Kaiserswerth was that it had developed organically in response to demand for social services and the provision of training. This differed, she noted in 1850, from the manner in which Victorian institutions were generally founded, with 'a castle in the air … a list of subscribers with some royal and noble names at the head [and] a double column of rules and regulations'.[53] The Establishment for Gentlewomen During Illness, which Nightingale took over in August 1853 and ran until leaving for the Crimean War in October 1854, was an example of this top-down method. It originated in discussions between two prominent ladies in the late 1840s: Janet Kay-Shuttleworth, the heiress wife of the doctor, politician, and proponent of residential institutions for schoolteachers, Sir James, and Charlotte Canning, an artist, courtier, and future vicereine of India (as wife of the viceroy Charles Canning). In October 1848 or 1849 Kay-Shuttleworth sent Canning 'a rough draft of a scheme which Lord Ashley & I have been concocting … and which we hope you will join in'. The proposal was 'to found an institution with the following objects':

> 1st. The admission of sick ladies of limited means at a moderate rate of payment.
>
> 2nd. The training of nurses of a superior class to attend on the sick.
>
> It is hoped that Ladies able to afford a higher rate of payment may avail themselves of the institution during illness. For this purpose distinct apartments will be provided.
>
> Admission to the 1st and 2nd departments of the Institution to be by a Committee of Ladies. The finance of the business to be conducted by a committee of Gentlemen.
>
> The staff will consist of Medical Officers to be non-resident. A resident superintending Lady who will also act as secretary, & resident nurses of good education to preside over the younger nurses & direct them in the care of the invalid inmates & in the course of their training.[54]

The institution opened on 15 March 1850, at 8 Chandos Street, Cavendish Square, London. Its patrons included Queen Victoria, Prince Albert, two Duchesses, and the Bishop of London, and its presiding council was made up of an assortment of aristocrats. Their support, combined with payments from patients, provided an annual income of approximately £1000.[55] Oversight was exercised by a Ladies' Committee, led by Canning and including Elizabeth Herbert, whose politician husband Sidney would oversee Nightingale's Crimean War mission.

The institution's focus on 'ladies of limited means' was of a piece with a wider trend to offer charity and support to women of the governess class in this period. Due to the oversupply of labour in governessing and dressmaking, the lack of other professions open to these women, and an apparent shortfall of men highlighted in the 1851 census, these women were seen to be at risk of becoming, in contemporary terminology, 'redundant', 'surplus', or 'superfluous'.[56] The desire to provide charitable support to such women—as opposed to, say, the far more numerous working-class women in poverty—has been explained in terms of an anxious identification on the part of middle- and upper-class Victorians with the poorest members of the gentry, who were at risk of losing respectable status altogether.[57] Harley Street, where the Establishment became based from August 1853, was also the location of the Governesses' Benevolent Institution (GBI), a charity founded in 1841 'offering assistance privately and delicately to ladies in temporary distress'.[58] In 1848, Queen's College, a pioneering institution for the education of governesses and girls, was set up next door to the GBI, with Kay-Shuttleworth and Canning among the lady visitors who attended lectures to ensure that propriety was upheld.[59] The Establishment for Gentlewomen During Illness was therefore an addition to a cluster of newly formed institutions dedicated to supporting governess-class women.

The *raison d'être* of the Establishment was, according to its secretary William Spring Rice, to offer health care superior to that available to these women (who were generally too poor to be admitted to the voluntary hospitals) in their own homes.[60] Setting up a small, private-yet-subsidised hospital establishment would offer such women 'the benefits of a Public Institution combined with the comforts and privacy of a home'. They would receive 'that sympathy, the absence of which is the most trying evil of illness in a lonely lodging'. They would, in other words, get the best of both worlds: the medical and nursing skills available in hospitals, and the atmosphere and 'comforts' of a 'home'.[61] Rice was keen to stress that the Establishment would enhance and complement, rather than interfere with, the moral benefits of middle-class domesticity.

The Establishment's managerial position for a 'resident superintending Lady' was a hybrid role, requiring skills in nursing and patient care, but also in household management. It was not easy to find women combining these skills. The first two appointments, a Miss Hall and a Miss Woolley, both experienced nurses, were quickly dismissed. Hall was 'not altogether equal to the … complicated duties of organising', while Woolley was accused of being 'neglectful and unkind to the patients'.[62] Given these failures, Nightingale's highly unusual combination of elite social background, organisational skills, desire to work, and commitment to and training in nursing appealed strongly to the Ladies' Committee. Nightingale reported to her family that they were so keen to have her that they offered her '"the appointment of the chaplain", the "dictatorship of the funds", the "choice of the house", "everything" but the admission of patients, *if* I would but come'.[63] Though Nightingale later wrote that some Committee members had objected to her 'youth and inexperience', she was nonetheless able to insist on taking the post 'without conditions', meaning that 'I shall undertake nothing without a clear understanding that I am to be left perfectly free, that I am to organise the thing and not they'.[64] Intentionally or not, Nightingale thus appropriated to herself the position of ultimate authority held at Kaiserswerth by the male pastor—and made sure to choose a quiescent chaplain.[65]

The attraction of the post for Nightingale lay in the Committee's second stated aim, to turn the Establishment into a site for the training of nurses 'of a superior class'. It had made no progress towards this goal by 1853, instead seconding a trained nurse each from Elizabeth Fry's institute and St John's House. Nightingale's ideas on training at this time were laid out in a set of 'rules for a Protestant nursing order' that she drafted in February 1853—with references to 'Chandos Street' and 'the committee' indicating that she had the Establishment in mind.[66] This text shows the strong influence of Kaiserswerth on Nightingale's thinking, but with important modifications. She envisioned, as at Kaiserswerth, a two-headed structure, but instead of 'father' and 'mother' figures, she proposed two women: a (paid) superintendent in charge of the nursing, 'the entire housekeeping, accounts, direction of the sisters' time and material guardianship', and a (self-funding) 'mother' in charge of 'moral guardianship'. The twin structure would allow each woman to take an annual three-month recuperative break. The male chaplain would be an important presence but clearly not in overall charge. Similarly, the 'Surgeon must never be master of the institution'—the women would be in control. Wards would be limited to four patients, as at Kaiserswerth, and the patients would be 'treated as guests and not as prisoners'. Aspirant nurses would pay for their training, but unlike at Kaiserswerth, would have a housemaid to assist them with cooking and cleaning. Trainees would 'receive lessons in reading, writing, arithmetic, needlework, scriptures'. Each would have to 'write a short account of her own life for the "mother"', just as Nightingale had done for Caroline Fliedner. In terms of living arrangements, each sister would have 'a den to herself, however small', but all would 'dine together, with maids of all work, mother and superintendent'. Once qualified, the sisters would 'read scriptures, sing … and pray morning and evening in their own wards', 'note down in a book what they have read to show to chaplain', and 'keep in a book a weekly account of each patient's conduct and disposition'. Provision would be made for the sisters in their old age. The overall 'principle' would be 'that of a family'.[67]

This conception of a well-managed, well-trained order of nursing sisters, living as a 'family', represented Nightingale's long-term vision for the Establishment, which she anticipated would take several years to implement.[68] The problems facing her on assuming the post in summer

1853 were more immediate and practical. The first task was to find a new building to upgrade from the Chandos Street location. The house eventually chosen, in Harley Street, required significant renovation. Problems with workmen and gas supplies—Nightingale wanted a boiler on each floor, so nurses wouldn't have to traipse up and down for hot water—dogged her for months. She found that the furniture and linen were 'in a most dirty and neglected condition. The tablecloths, kitchen cloths, towels etc., were ragged ... The sheets were not good'.[69] Nightingale and her nurses—she did not hire a charwoman or needlewoman to help—thus spent a substantial part of the late summer months sewing: repairing sheets, pillowcases, tablecloths, dusters, toilet covers, towels, and much besides. Nightingale further found it 'desirable ... for the sake of securing economy and wholesome bread', for the Establishment to bake all its own bread and biscuits. She utilised her 'dictatorship of the funds' to change various grocery suppliers and contracts, procure furniture, and replace several household staff. She furnished more rooms to enlarge the Establishment's capacity to twenty-seven. This work preoccupied her to the extent that her November 1853 report to the governors, summarising her first three months in post, mentioned nurses and patients only in passing—though the patients soon became the focus of her thinking.[70]

In general, Nightingale believed in making patients feel like guests while in a care setting, and strongly upheld their rights to privacy and autonomy. In later years, Nightingale advised that nurses should assume the mantle of protectors of these rights, acting as 'the patients' defender and keeper'.[71] At Harley Street, Nightingale gave considerable thought to homely aspects such as the supply of good quality bedding, furniture, and food. She also encouraged the patients to feel more at home by coming out of their rooms, spending time in the communal drawing room, and asking visitors to read to them.[72] Nightingale's actions in this vein were typical of the nineteenth-century trend to import materials and behaviours from the genteel home into institutions, on the assumption that they would automatically be of moral and medical benefit. There is no available evidence as to what the patients made of this, though by summer 1854 the idea that Nightingale's Establishment was 'most admirably managed, the object to make it as much like a home as possible' had reached the artist Dante Gabriel Rossetti, then investigating treatment options for his troubled muse Lizzie Siddal.[73] As cautioned in Jane

Hamlett's work, however, it is important not to assume that these trappings of domesticity necessarily translated into experiences of homeliness for residents.[74]

Despite this initial emphasis on homely surroundings, Nightingale's long-term plans, and the economics of the Establishment, were reasons to ensure that the wrong kind of patient did not make themselves too much at home. Though patients were asked to contribute towards the cost of their care, the institution deliberately ran an operating loss, with charitable donations making up the shortfall.[75] This *modus operandi* created an incentive not to retain patients for too long, especially if their condition was not improving, so as to put the charitable funds to the most efficient use. Furthermore, Nightingale's vision of transforming Harley Street into a nursing order or training school required a range of what she termed 'proper objects for medical and surgical treatment', in order to provide trainees with appropriately varied study material.[76]

From Nightingale's point of view in 1853–1854, the ideal patient was one that came in for surgery or some other distinctive but finite medical procedure, recovered fairly quickly, and did not stay around too long. Unsurprisingly, many patients did not fit this bill. The first five discharged during Nightingale's tenure included one with chronic rheumatism, which she considered 'unfit for [the] institution', and one that she intriguingly claimed was 'completely cured by the prospect of New Zealand (weakness)'.[77] This soon became a pattern: Nightingale, like most managers of Victorian charitable hospitals, did not want to keep patients with long-term conditions that would only improve slowly; nor did she want patients suffering from so-called nervous conditions for long periods. In November 1853, she produced a memorandum that asserted the following principles:

> I – That it is rare, if a case be not better at the end of two months, that it will be better at the end of four or of six months … II – That such cases, if accustomed to look upon this Institution as a home, have no motive to be better at the end of two months, but rather to be worse, as this Institution is the most cheap and comfortable lodging-house which they can find, with the luxury of taking medicine, & of sympathy besides. III – That, of the cases which come to this Institution, ninety per cent are simply hysterical cases, to which these remarks more peculiarly apply.[78]

Her conclusion was that no patient should be allowed to stay in the house for longer than two months, 'unless there be prospect either of improvement or of death'. If this rule were not implemented—she argued in her report of February 1854—'this will become not a hospital for the sick, but a hospital for incompatible tempers and for hysterical fancies … A hospital is good for the seriously ill alone – otherwise it becomes a lodging house where the nervous become more nervous, the foolish more foolish, the idle and selfish more selfish and idle'.[79] Over the following three months, she discharged three patients labelled 'hysterical, no improvement', and four more 'cured of complaints which never existed'.[80] In other cases, she intervened to find her patients long-term places in other institutions.[81] 'There have been admitted many cases which proved to be unworthy of the charity', she argued in her May report.[82]

There was probably a part of Nightingale that felt that she had not left country house life and her nervously hypochondriac sister behind merely to spend the rest of her days ministering to the over-dramatised complaints of underemployed gentlewomen. But she was also seeking to draw a distinction: a hospital was not a home, and should not be treated as such. The problem was that it was not clear which category—hospital or home—Harley Street fitted into. It had hospital-like elements, with some patients admitted for short-term surgical or medical procedures. Nightingale wanted the Establishment to prioritise this aspect—she attended every operation that took place there—to assist her in establishing a nursing training institute.[83] Many of the patients (or their families), however, saw the Establishment as a residential care home. Thus, a significant proportion of its residents were women who were not in need of hospital care, but 'who have wearied out their families, or been wearied out by them', or simply those 'who have no families at all'.[84] Nightingale seems to have wanted the Establishment to cease catering to such women.

A third category were those who were undergoing extended recovery from surgery or illness and 'anxious to return to their families and save them expense'.[85] These too did not belong in a hospital environment, Nightingale felt. By the time she came to write *Notes on Hospitals* (1859), she had identified a need for a new kind of institution—convalescent homes—to solve this problem. Starting from the principle that 'no patient ought ever to stay a day longer in hospital than is absolutely essential for medical or surgical treatment', Nightingale declared that 'every hospital should have its convalescent branch, and every county its convalescent home'.[86] This would benefit patients as well as hospitals, since 'as long as they are hospital inmates, they feel as hospital inmates, they think as hospital inmates, they act as hospital inmates, not as people recovering'. Convalescent houses could be decorated in a more homely way than hospitals, and patients could move around more, take on odd jobs, or do gardening. Moving to a convalescent home would 'get rid of the idea of being in hospital altogether from the minds of the inmates', and 'substitute for it that of home'.[87] Although Nightingale believed that strict supervision and gender segregation would be needed to maintain moral discipline in convalescent homes, the focus of convalescence was on creating home-like environments to aid recovery, with home seen both as possessing intrinsic healing qualities and as the patients' final goal and destination (Fig. 5.2).

In 1854, with her ideas on convalescence not yet fully formed, Nightingale simply attempted to exclude non-medical patients from Harley Street. This had an inevitable impact on numbers, such that by May only a third of the bedrooms were occupied—thereby reducing revenue and increasing per-resident costs.[88] She asked the Ladies' Committee to turn its 'immediate attention' to the 'deficiency of patients' lest the institution 'degenerate into a luxurious piece of charity'.[89] The strength of her language perhaps stemmed from frustration at the realisation that, given the lack of medical patients and training infrastructure, she was not going to be able to fulfil her ambitions for a nursing order.[90] Well before she submitted her formal resignation (giving six months' notice) in August, she had begun to look elsewhere. In her mind, establishing the 'order and system' that Canning praised in her reply to Nightingale's resignation had only been a preliminary step on the way to a much larger reform of nursing.[91] Her ambitions had outgrown the Establishment and required a broader canvas. Had events in the Crimea not intervened, she may well have taken up an offer to assume the superintendence of the nursing at King's College Hospital (whose lead surgeon, William Bowman, also attended Harley Street).[92]

Nightingale's Harley Street experience shows how far her ideas of home were bound up with her ideas on women and work. She was content in her early months there to take on the conventional housemistress or matronly task of household management and efficiency, happily undertaking sewing, cooking, or shopping work as required. Just doing the work, after all her struggles with her family, must have felt like an achievement in itself—as the pioneering woman doctor Elizabeth Blackwell wrote to her, simply by 'weakening the barriers of prejudice

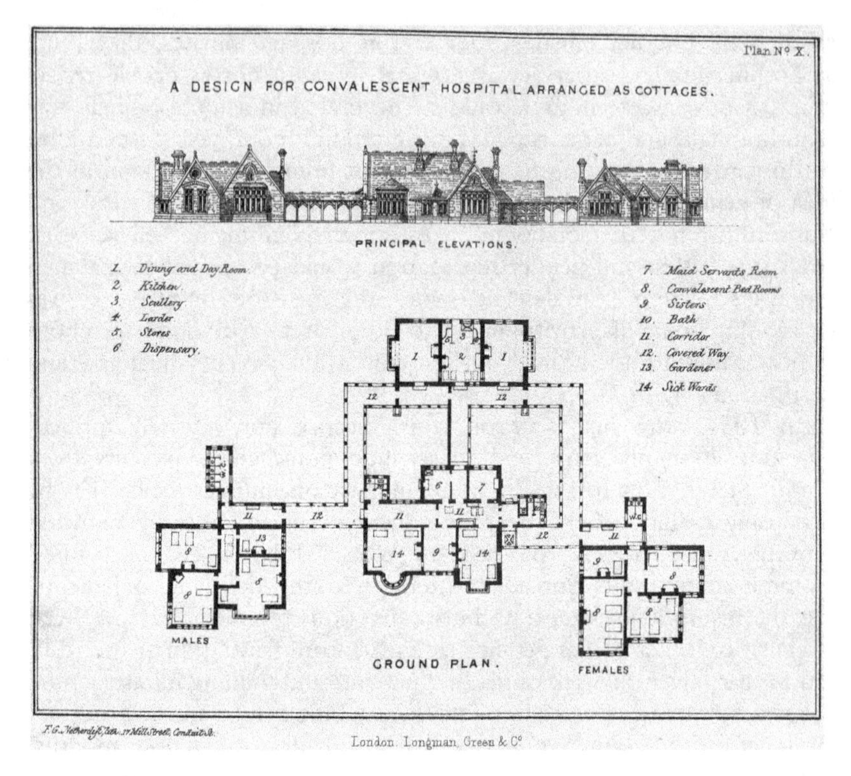

Fig. 5.2 Sketch of a convalescent hospital arranged as cottages, F. G. Netherdift, in Florence Nightingale, *Notes on Hospitals* (London: Longman, 1863; first published by John Parker in 1859 without this plan), 112

which hedge in *all* work for women – you thus carry out a reform, much wider than the ostensible nursing plan'.[93] As time went on, however, Nightingale came to chafe against the Establishment's limitations and to resist the possibility that she and any nursing trainees would primarily become long-term carers to residents with chronic conditions. She would not allow such patients to feel at home in her institution if the price was that she could not feel at home in her work.

Home Sisters, Mother-Chief: The Nightingale School and Nightingale Home

Five years passed before Nightingale was able to resume plans for a nurse training scheme for civilian hospitals. Much had changed in that time. Nightingale's post-Crimea celebrity ensured that any new nursing venture she undertook would attract national attention, but it was no longer her top priority, and being practically housebound with brucellosis from 1856 (Chapter 7), she was no longer in a position to take personal charge. She spent much of the 1856–1860 period working towards improving the health and sanitation of the British army, as well as writing *Notes on Nursing* and *Notes on Hospitals*. Her post-war writings on trained nursing concerned military rather than civilian hospitals. It was only in late 1859 and 1860 that she turned her attention to what became the Nightingale School.

During the Crimean War, the British public had conferred upon Nightingale a substantial pot of money to put towards nurse training and gave Nightingale herself the responsibility for spending it. The Nightingale Fund, created following a London public meeting held in her honour in November 1855 and overseen by a committee of eminent figures including Sidney Herbert and Richard Monckton Milnes, raised £44,039—several million in today's money—by the time of its closure in June 1856.[94] This fund represented an important opportunity, both due to its size and because its high profile helped spread the idea that trained and skilled nursing was a respectable and desirable occupation for women. However, it came with strings attached. Though most of the money came from wealthy donors, the Fund was also arguably 'the first national appeal aimed at all classes', with 20 percent of contributions coming from soldiers.[95] This wide range of support dictated that the Fund had to be directed towards secular ends: donations from northern nonconformists and Irish Catholics (over-represented in the army) could not be used to create an Anglican order along the lines of St John's House, nor even the non-denominational Protestant order envisaged by Nightingale in 1853—at least not without courting political controversy. Another constraint was that the Fund Council, taking a long-term view, chose not to spend its capital, only its investment returns, which generated an annual income of approximately £1500.[96] While substantial, this was in the same ballpark as Harley Street's budget—not enough to establish a major new hospital or institution.[97] As a result, the Fund would need to work with and through others in order to achieve large-scale change.

The Crimean War had reinforced Nightingale's belief that clinical experience, preferably in a large teaching hospital, was essential to developing real nursing expertise and professional judgement.[98] To provide nurses with such experience through their training, the Fund decided to seek a partnership with a London teaching hospital, which at the time were the country's most prestigious medical institutions. The choice of St Thomas' was influenced by Nightingale's good relationships with the Matron, Sarah Wardroper, whom she had met in 1854, and the Medical Officer, Richard Whitfield, who supported the scheme strongly at a time when many doctors were opposed.[99] Obtaining the School was a coup for St Thomas', allowing it to benefit from the prestige of Nightingale's name and the labour of highly motivated nursing trainees inspired by her Crimean example. Its governors managed to negotiate terms that were highly favourable to the hospital, allowing it to make money from providing board and lodging to the trainees and tying the Fund into providing large numbers of trainees over a seven-year contract.[100]

Nightingale consulted Mary Jones, the head of St John's House and nursing superintendent at King's College Hospital, on these arrangements. By 1860 Jones had far more experience of civilian hospitals and nurse training schemes than Nightingale, and the two would subsequently work together on a scheme for training midwifery nurses at King's, supported by the Nightingale Fund.[101] The thrust of Jones's advice was that Nightingale should retain as much independence as possible, and not enter into a one-sided contract that prioritised the hospital's needs—the hospital authorities, Jones wrote to Nightingale, displayed 'every disposition to drive a hard bargain … and to make the whole matter a paying business to themselves'.[102] She suggested in March 1860 that the number of trainees be limited to fifteen; if the Fund wanted to train more, it should 'make some similar arrangement with another Hospital' and set up its own independent Home 'in some central position'.[103] Jones also suggested that the Fund should commit for a much shorter period than seven years. Even though Nightingale followed much of this advice, in 1866 *The Lancet* still expressed shock on discovering just how much money the hospital was making out of its arrangement with the Fund.[104]

Jones's emphasis on not ceding too much control to the hospital was a reminder of the difficulty women nursing reformers had in retaining autonomy at a time when neither hospital authorities nor doctors were unanimously in favour of an independent, trained nursing service.[105] Setting up her School, Nightingale insisted that nurse trainees must not be answerable to male doctors or chaplains. As she wrote to Jones in 1867, '[t]he whole reform in Nursing both at home and abroad has consisted in this: to take all power over the Nursing out of the hands of the men and put it into the hands of *one female trained* head ... Women are never governed by a man, except to their own detriment'.[106] But since the housebound Nightingale could not herself be that authority, this meant entrusting crucial decisions and responsibilities regarding the Nightingale School to the Matron, Wardroper, who was not a nurse, and who answered first and foremost to the hospital, not the Fund. The training itself would be conducted by Wardroper's ward sisters, many of whom were not versed in the methods of reformers like Nightingale and Jones, and who had no teaching qualifications or training.

Nonetheless, for the first decade of the School's operation, Wardroper, Nightingale, and her cousin Henry Bonham-Carter, Fund Secretary from 1862, formed a governing triumvirate who agreed on most matters. Though Nightingale only ever visited the School once (in 1882), she kept abreast of all the probationers and their progress, wrote them recommendations for future roles, and maintained extensive contact with them after they graduated. The School was inundated with demands from hospitals and other institutions to supply trained nurses. In 1864, the philanthropist William Rathbone funded extra accommodation for increased nurse training places at the Nightingale School in order to train nurses for a workhouse infirmary in Liverpool, to which Agnes Jones and a team of nurses were sent in 1865.[107] In 1865, the School introduced 'Specials', 'lady' probationers who paid for their board and expenses (though not for their training), a measure which increased the number of nurses trained for senior administrative roles, at the price of embedding class differences within the School. The scheme produced senior nurses such as Elizabeth Torrance, who took over the nursing at Highgate Infirmary, and Angelique Lucille Pringle, superintendent at Edinburgh Royal Infirmary from 1874 to 1887, where she established a nurse training institute before returning to St Thomas' to succeed Wardroper as Matron.[108] The Nightingale School thus rapidly became a motherhouse in its own right, sending teams out across the UK as well as to Sydney, Australia and Montreal, Canada.[109]

By the early 1870s, however, Wardroper's methods and personnel decisions were being called into question by probationers and former students, who reported their dissatisfaction to Nightingale. The Fund also became profoundly dissatisfied with its lecturer, the Medical Officer Richard Whitfield, whose alcoholism and petty corruption had become conspicuous.[110] When the Hospital relocated to its new site on Albert Embankment in 1871, Whitfield was induced to resign. Wardroper remained, but the Fund obtained the right to at least be consulted on her eventual replacement.

An important change accompanying the Albert Embankment move was the construction of a dedicated residence for the trainees: the Nightingale Home. From the beginning of the School, Nightingale paid considerable attention to the question of accommodation, believing that nurses' status as professional people, deserving of respect, privacy, and autonomy, should be reflected in the 'space and comfort' of their living conditions.[111] When the School had opened in 1860, the trainees were lodged in a former ward on the attic floor of an old St Thomas' building in Southwark. There was a dining room, sitting room, and a separate cubicle or 'compartment' for each probationer, though the partitions did not reach the ceiling. Nightingale did her best to make the place cheerful and welcoming by ensuring that prints, books, and flowers were supplied.[112] Between 1862 and 1871, the hospital was temporarily based in Surrey Gardens. On her arrival Agnes Jones, later one of the School's star pupils, described:

> a large, lofty, comfortable room, with tables, chairs, flowers, pictures, books, carpet, rug, fire, gas, like any sitting-room: off this, surrounded by the varnished boards, are the little bedroom cells, their wooden walls about ten feet high, not half way to the ceiling, with a bed, small chest of drawers, wash-stand, chair and towel-rail … I am now writing at one of the numerous little tables, with bright flowers and numbers of all kind of magazines around me … There is a temporary church fitted up in the house … Each cell has its own gas … [there is also a] nice light eating-room, quite separate from, but near our sitting-room … I feel at home already.[113]

This extract suggests that the School had successfully imported enough of the trappings of middle-class domesticity, combined with the promise of useful work, to put at least some new arrivals at their ease. However, other trainees felt that the Surrey Gardens lodgings were cramped, and at one point the School reduced the number of trainees by five owing to accommodation problems.[114]

The establishment of the Nightingale Home in 1871 was thus the first opportunity to create purpose-built accommodation. The Home contained bedrooms for thirty-eight probationers and had been carefully planned as an integral part of the new hospital's design, on which Nightingale had been consulted.[115] By housing the trainees together, close to the Matron but away from the wards, it began to give physical embodiment to Nightingale's idea that 'ward training is but half of training', with the other half consisting of inculcating 'habits of order, cleanliness, regularity and moral discipline' as well as theoretical knowledge.[116] The Home would safeguard the probationers' respectability, away from the moral and sanitary dangers of the wards, and keep them in a disciplined environment under the Matron's supervision, hopefully facilitating the kind of character training and *esprit de corps* that Nightingale had found so valuable at Kaiserswerth. In 1875, Nightingale defined the Home as 'a place of moral, religious and practical training; a place of training of character, habits, intelligence, and a place of acquiring knowledge – technical and practical'.[117] It became a key part of what was understood as the system of Nightingale nursing: when other hospitals sought to imitate its model, the Fund insisted on the incorporation of a proper Nurses' Home, despite the expense and difficulty this could add.

Just as important as the institution of the Nightingale Home was the creation of the new post of Home Sister to supervise it. This was not in fact Nightingale's idea, but that of Elizabeth Torrance, the former Special Probationer who had gone on to superintend the nursing at Highgate infirmary. In April 1872, Torrance wrote to Nightingale outlining the 'very great need of someone to take charge of' or 'settle down' the trainees in the Home and proposing her own suitability for 'such a post as Home Sister, or Training Sister'.[118] As developed in discussion with Nightingale, the Home Sister became something of a hybrid position, combining responsibility for academic lessons in the Home (as opposed to practical training on the wards) with pastoral duties of moral guardianship and the maintenance of discipline. However, it represented an obvious reduction of Wardroper's authority. Wardroper became something of an obstacle for Torrance and her successors, preventing the full integration of theoretical and practical training by refusing to allow the Home Sister to attend ward classes.[119] Nonetheless, the Home Sister became a pivotal figure in the School—especially Maria Machin (Home Sister 1873–1875), whom Nightingale credited with making the Home into 'what we have been always craving for', and then Mary Crossland (in post 1875–1896), who worked effectively with the new Medical Officer, Dr John Croft, to produce a coherent academic curriculum.[120] As part of her moral guidance role, Crossland also led voluntary Bible classes on Sunday afternoons, which most trainees attended, and organised some singing lessons and the occasional cookery class.[121]

The system that the School employed after 1872 therefore resembled the two-headed structure that Nightingale had envisaged back in 1853 of Matron-superintendent and mother. Except that there were three heads, with Nightingale herself now devoting greater time and emotional energy to the School, and eventually referring to herself as its mother-chief. She considered moving from South Street, Mayfair, where she had lived since 1865, to be closer to the Home, but was dissuaded on the grounds that this might seem to undermine Wardroper's authority.[122] Instead, she encouraged the probationers to pay her regular visits, during which she made careful notes to inform her subsequent conversations with Wardroper and Crossland. She began to write addresses to the probationers, which were read out at the Home's summer parties, and organised an annual outing to Claydon House, Parthenope's home in Buckinghamshire. These addresses emphasised the personal, spiritual, and moral qualities required of nurses, exhorting them to be 'trustworthy, punctual, quiet and orderly, cleanly and neat, patient, cheerful, and kindly' and to search for 'greater individual responsibility, greater self-command… greater self-possession'.[123] She compared nurses to 'missionaries' spreading the 'virtues of civilisation', but also reminded them that 'this is not individual work … remember we are not so many small selves, but members of a community'.[124] That 'community' was also to be a family. 'My dear Children', began Nightingale's last 'address', in 1900, 'you have called me your Mother-chief, it is an honour, to call you my children'.[125] Thus in another manifestation of her complex relationship with the ideology of home, Nightingale's active life, which began with her stark challenge to conventional upper-class gender roles, ended with her appropriating the language of family to serve the social motherhood role she took up among her trainees.

'A Real Home, Within Reach of Their Work, for the Nurses to Live In'

When Nightingale's parents refused to allow her to go to Salisbury Hospital to work as a nurse in 1845—because to do so would be physically dangerous and damaging to her respectability—they in effect presented a challenge. That challenge was of bringing nursing into the fold of respectable society in ways that would alleviate the problem of educated women's under-employment. Changes to nurses' living arrangements were central to this process. First, nurses moved from having to live and wash on the wards into separate, secure living environments where respectability was guaranteed by strict discipline. Sisterhood houses, and later Nurses' Homes (with the Nightingale Home at St Thomas' as the prototype), were moral and upright spaces, but they also increasingly became comfortable, too, offering homely features such as sitting rooms and flowers that encouraged a sense of well-being and self-respect among their occupants. Uniforms and physical separation from doctors and patients emphasised that nurses were no longer to be fair game for sexual harassment and exploitation. Second, the trainee nurses' living arrangements were secularised and feminised by removing ultimate authority from male chaplains and placing it in the hands of a female superintendent—later supplemented by a Home Sister who added pastoral and educational duties to those of household management. Nightingale Homes, under the exclusive control of a female hierarchy, were to be places of education, discipline, privacy, and retreat. Curiously, perhaps, it was only in later decades that these principles were extended to all nurses, as opposed to only trainees—Gassiot House, the dedicated Nurses' Home at St Thomas' hospital, did not open until 1906.

These changes were suffused with language and concepts drawn from the ideology of the bourgeois home. Family terminology and metaphors inflected the nineteenth-century nursing workplace, even if there was also a period, especially in the years following the Crimean War, when it was affected by the militarised language of discipline, battles, and sacrifice.[126] The deployment of the terminology of home and family—by the religious sisterhoods, by the Fliedners at Kaiserswerth, by Nightingale and others—was consciously chosen in order to help confer social respectability onto their projects, reassuring the wider world (and, not least, the parents of potential nurses) of the safety and virtue of their efforts. At the same time, home denoted an ideal of communal endeavour and commitment that Nightingale held dear, and which challenged the class- and gender-bound limitations of the period. In this sense, home was an ideal closely linked to the noble pursuit of work; homes for nurses were places in which, as Nightingale wrote in 1876, 'good young women' could 'live and nurse'. The disillusionment that Nightingale felt towards the conventional family inspired her to create surrogates in the institutions over which she presided; she wanted to establish and support 'a real home, within reach of their work, for the nurses to live in – a home which gives what real family homes are supposed to give'.[127] Such wide-ranging deployments of the concept of home began to shift the meaning of the term itself. Home could now refer to a base for the training of a community of professional women, a meaning that could not have been applied in the eighteenth century.

Nightingale did not initiate this process. She studied and borrowed much from the Fliedners and from the sisterhoods, but she also adapted and innovated. Her insistence on a female-only hierarchy for nurses and the principle of their absolute independence from doctors had long-term significance, as did her stress on the importance of the Nurses' Home as a guarantor of minimum standards in living arrangements, such as a separate bedroom for each trainee.

The Nightingale School was not a perfect system, nor was it a pure expression of Nightingale's vision. It was, rather, a pragmatic compromise that balanced the needs of St Thomas' management, the constraints of the Nightingale Fund, and the societal demand to produce trained nurses quickly. But it was an important innovation, a model that was soon replicated in hospitals across Britain and around the world in a way that the sisterhoods had not been able to achieve: a true motherhouse, in the Kaiserswerth sense. In the process, the Nightingale School made a substantial contribution to ensuring that, by the late nineteenth century, there were many more homes available to young women seeking paid, rewarding, and respected work in health care.

Notes

1. Elizabeth Gaskell to Emily Shaen, 27 October 1854, in *The Letters of Mrs Gaskell*, ed. J. A. V. Chapple and Arthur Pollard (Manchester: Manchester University Press, 1966), 320.
2. The idea of nineteenth-century institutions as vectors of power and discipline was argued most influentially by Michel Foucault, in e.g. *Discipline and Punish: The Birth of the Prison*, trans. Michael Sheridan (New York: Pantheon, 1977).
3. See Patrick Joyce, *The State of Freedom: A Social History of the British State Since 1800* (Cambridge: Cambridge University Press, 2013), 6.

4. Nightingale, 'Hospital Construction: Wards', *The Builder* 16,816, 25 September 1858, 641–643, in *The Collected Works of Florence Nightingale, Volume 16: Florence Nightingale and Hospital Reform*, ed. Lynn McDonald (Waterloo, ON: Wilfrid Laurier University Press, 2013) [hereafter *CW* 16], 307.
5. Jane Hamlett, *At Home in the Institution: Material Life in Asylums, Lodging Houses and Schools in Victorian and Edwardian England* (Basingstoke: Palgrave Macmillan, 2015).
6. Martha Vicinus, *Independent Women: Work and Community for Single Women, 1850–1920* (Chicago: University of Chicago Press, 1985), 1–6.
7. Ellen Jordan, *The Women's Movement and Women's Employment in Nineteenth Century Britain* (London: Routledge, 1999).
8. Sue Hawkins, *Nursing and Women's Labour in the Nineteenth Century: The Quest for Independence* (Abingdon: Routledge, 2010).
9. Eva Gamarnikow, 'Nurse or Woman: Gender and Professionalism in Reformed Nursing, 1860–1923', in *Anthropology and Nursing*, ed. Pat Holden and Jenny Littleworth (London: Routledge, 1991), 110–129.
10. Carol Helmstadter and Judith Godden, *Nursing Before Nightingale, 1815–1899* (Farnham: Ashgate, 2011).
11. Ibid., 24, 48.
12. Nightingale, *The Institution of Kaiserswerth on the Rhine, for the Practical Training of Deaconesses* (London: London Ragged Colonial Training School, 1851), in *The Collected Works of Florence Nightingale, Volume 7: Florence Nightingale's European Travels*, ed. Lynn McDonald (Waterloo, ON: Wilfrid Laurier University Press, 2004) [hereafter *CW* 7], 492–511, 499.
13. Nightingale to Elizabeth Herbert, 29 May 1854, in *The Collected Works of Florence Nightingale, Volume 8: Florence Nightingale on Women, Medicine, Midwifery and Prostitution*, ed. Lynn McDonald (Waterloo, ON: Wilfrid Laurier University Press, 2005), 644.
14. Helmstadter and Godden, *Nursing Before Nightingale*, 125–168.
15. Vicinus, *Independent Women*, 46–56.
16. Ibid., 54.
17. Nightingale to Hilary Bonham-Carter, 11 December 1845, in Edward Cook, *The Life of Florence Nightingale*, vol. 1 (London: Macmillan, 1913), 44.
18. The full register of 'Nurses sent to the Military Hospitals in the East' is held at the Florence Nightingale Museum.
19. See F. F. Cartwright, 'Miss Nightingale's Dearest Friend', *Proceedings of the Royal Society of Medicine* 69/3 (March 1976): 169–175.
20. Cook, *Life*, vol. 1, 64, citing a now-lost Nightingale diary for 7 October 1846.

21. Mark Bostridge, *Florence Nightingale: The Woman and Her Legend* (London: Penguin, 2009), 85.
22. Nightingale note of 15 June 1848, cited in Cook, *Life*, vol. 1, 110.
23. R. G. Huntsman, Mary Bruin and Deborah Holttum, 'Twixt Candle and Lamp: The Contribution of Elizabeth Fry and the Institution of Nursing Sisters to Nursing Reform', *Medical History* 46 (2002): 351–380.
24. Uwe Kaminsky, 'German "Home Mission" Abroad: The *Orientarbeit* of the Deaconess Institution Kaiserswerth in the Ottoman Empire', in *New Faith in Ancient Lands: Western Mission in the Middle East in the Nineteenth and Early Twentieth Centuries*, ed. Heleen Murre-van den Berg (Leiden: Brill, 2006), 191–210, 191.
25. Karen Nolte, 'Protestant Nursing Care in Germany in the 19th Century: Concepts and Social Practice', in *Routledge Handbook on the Global History of Nursing*, ed. Patricia D'Antonio, Julie A. Fairman, and Jean C. Whelan (Abingdon: Routledge, 2013), 167–182.
26. Nightingale's 1851 diary and notes, *CW* 7, 517–584.
27. Nightingale, note, 15 June 1848, *CW* 7, 489.
28. Selina Bracebridge to Nightingale, 3 August 1850, *CW* 7, 490–491.
29. Ibid.
30. Nightingale to her father, 15 August 1850, in *The Collected Works of Florence Nightingale, Volume 1: Florence Nightingale: An Introduction to Her Life and Family*, ed. Lynn McDonald (Waterloo, ON: Wilfrid Laurier University Press, 2001) [hereafter *CW* 1], 232.
31. Helmstadter and Godden, *Nursing Before Nightingale*, 70.
32. Nightingale, *Institution of Kaiserswerth*, *CW* 7, 500.
33. Ibid., 504.
34. Nightingale to Henry Bonham-Carter, 24 July 1867, *CW* 7, 598.
35. Nightingale's 1851 Kaiserswerth diary, *CW* 7, 517–543.
36. Nightingale, *Institution of Kaiserswerth*, *CW* 7, 499–502.
37. Ibid., 502.
38. Nightingale to her father, 15 August 1850, *CW* 1, 232; *Institution of Kaiserswerth*, *CW* 7, 499–502.
39. Bostridge, *Florence Nightingale*, 144.
40. Eileen Janes Yeo, 'Social Motherhood and the Sexual Communion of Labour in British Social Science, 1850–1950', *Women's History Review* 1, no. 1 (1992): 63–87, 75.
41. Nightingale, *Institution of Kaiserswerth*, *CW* 7, 502.
42. Nightingale to her father, 15 August 1850, *CW* 1, 232.
43. Lynn McDonald, 'Kaiserswerth, 1850', *CW* 7, 490.
44. Nightingale's diary, *CW* 7, 519, 526.
45. Lynn McDonald, 'Nightingale's Preparation for Nursing', *The Collected Works of Florence Nightingale, Volume 12: The Nightingale School*, ed. Lynn McDonald (Waterloo, ON: Wilfrid Laurier University Press, 2009) [hereafter *CW* 12], 51.

46. Nolte, 'Protestant Nursing Care', 173–174.
47. Nightingale to her mother, 16 July 1851, *CW* 1, 127.
48. Nightingale's diary, 8 August 1851, *CW* 7, 528.
49. Nightingale's diary, 19 August 1851, *CW* 7, 532.
50. Nightingale to her father, 15 August 1850, *CW* 1, 232.
51. Nightingale, *Institution of Kaiserswerth*, *CW* 7, 504–505.
52. Nightingale, *Suggestions for Thought*, in *The Collected Works of Florence Nightingale, Volume 11: Florence Nightingale's Suggestions for Thought*, ed. Lynn McDonald (Waterloo, ON: Wilfrid Laurier University Press, 2008), 508.
53. Nightingale, *Institution of Kaiserswerth*, *CW* 7, 499.
54. 'Report of the Establishment for Gentlewomen During Illness', March 1852, BL Canning papers, Add. Mss Eur F699/2/3/1.
55. Ibid.
56. Kathrin Levitan, 'Redundancy, the "Surplus Woman" Problem, and the British Census, 1851–1861', *Women's History* Review 17, no. 3 (2008): 359–376.
57. Jordan, *The Women's Movement*, 32, 62; Mary Poovey, *Uneven Developments: The Ideological Work of Gender in Mid-Victorian England* (Chicago: University of Chicago Press, 1988), 127.
58. Cited in Jordan, *The Women's Movement*, 115.
59. Ibid., 117.
60. 'Report of the Establishment for Gentlewomen During Illness', March 1852.
61. Ibid.
62. Letter book of the Ladies' Committee of the Establishment for Gentlewomen During Illness, Florence Nightingale Museum, TN050.
63. Nightingale to her parents, 8 April 1853, *CW* 12, 63.
64. Nightingale to her father, 30 August 1853, *CW* 12, 80.
65. Helmstadter and Godden, *Nursing Before Nightingale*, 80.
66. Nightingale, 'Draft Rules for a Protestant Nursing Order', 14 February 1853, *CW* 12, 66–70, 66, 67.
67. Ibid., 66–69.
68. Nightingale to Theodore Fliedner, 10 September 1853, Wellcome 9083/3, cited in Bostridge, *Florence Nightingale*, 194.
69. Nightingale's first 'Quarterly Report to the Governors', 14 November 1853, *CW* 12, 99–104.
70. Ibid.
71. Nightingale to Marie von Miller, 17 March 1879, in *The Collected Works of Florence Nightingale, Volume 13: Florence Nightingale: Extending Nursing*, ed. Lynn McDonald (Waterloo, ON: Wilfrid Laurier University Press, 2009), 475.

72. Unknown sender ('Puffin') to Parthenope Nightingale, 1 November 1853, Wellcome 9048/9: 'I read to them in the afternoon & in the evening ... Flo likes me to get them to come to the drawing room and be read to.'
73. Cited in Bostridge, *Florence Nightingale*, 195.
74. Hamlett, *At Home in the Institution*, 1–12.
75. 'Report of the Establishment for Gentlewomen During Illness', March 1852. In 1851 patient contributions amounted to £437, compared to expenditure of £1098.
76. Nightingale's fourth 'Quarterly Report to the Governors' of Harley Street, 7 August 1854, *CW* 12, 112.
77. Nightingale's first 'Quarterly Report', *CW* 12, 103.
78. Nightingale, memorandum dated 25 November 1853, Claydon N371, *CW* 12, 85.
79. Nightingale's second 'Quarterly Report', 20 February 1854, *CW* 12, 104–107, here 106, 105.
80. Nightingale's third 'Quarterly Report', 15 May 1854, *CW* 12, 107–111.
81. Nightingale to unnamed recipient, 24 February 1854, *CW* 12, 89–90.
82. Nightingale's third 'Quarterly Report'.
83. Gaskell, *Letters*, 306.
84. Nightingale's second 'Quarterly Report', *CW* 12, 105.
85. Ibid.
86. Nightingale, *Notes on Hospitals*, in *CW* 16, 164–171, 164.
87. Ibid.
88. Nightingale's third 'Quarterly Report', *CW* 12, 110.
89. Ibid.
90. Bostridge, *Florence Nightingale*, 197.
91. Canning to Nightingale, 8 August 1854, *CW* 12, 97.
92. Bostridge, *Florence Nightingale*, 197–198.
93. Elizabeth Blackwell to Nightingale, 27 March 1854, Claydon N370.
94. Monica Baly, *Florence Nightingale and the Nursing Legacy* (Beckenham: Croom Helm, 1986), 16–17.
95. Ibid., 17; Roy Wake, *The Nightingale Training School 1860–1996* (London: Haggerston, 1998), 45.
96. Baly, *Nursing Legacy,* 171.
97. Ibid., 45. Income was £1426 in 1862, equating to a 3 percent annual investment return on the £45,000 Fund.
98. Helmstadter and Godden, *Nursing Before Nightingale*, 121.
99. Lucy Seymer, *Florence Nightingale's Nurses: The Nightingale Training School* (London: Pitman, 1960), 27–28.
100. Baly, *Nursing Legacy*, 50–51.

101. See *The Collected Works of Florence Nightingale, Volume 8: Florence Nightingale on Women, Medicine, Midwifery and Prostitution*, ed. Lynn McDonald (Waterloo, ON: Wilfrid Laurier University Press, 2005), 141–408.
102. Mary Jones to Nightingale, 10 May 1860, BL Add Mss 47743 ff9–10.
103. Jones to Nightingale, 28 March 1860, BL Add Mss 47743 ff3–6.
104. *The Lancet*, 31 March 1866.
105. See Judith Moore, *A Zeal for Responsibility: The Struggle for Professional Nursing in Victorian England, 1868–1883* (Athens: University of Georgia Press, 1988).
106. Nightingale to Mary Jones, 8 January 1867, in *The Collected Works of Florence Nightingale, Volume 3: Florence Nightingale's Theology: Essays, Letters and Journal Notes*, ed. Lynn McDonald (Waterloo, ON: Wilfrid Laurier University Press, 2002), 468–469.
107. Baly, *Nursing Legacy*, 42.
108. On Pringle see Lynn McDonald's biographical sketch, *CW* 12, 887–890.
109. Judith Godden, 'The Dream of Nursing the Empire', in *Notes on Nightingale: The Influence and Legacy of a Nursing Icon*, ed. Sioban Nelson and Anne Marie Rafferty (Ithaca: Cornell University Press, 2010), 55–75.
110. Baly, *Nursing Legacy*, 150–157.
111. See Lynn McDonald, 'Introduction' to *CW* 16, 18ff.
112. Seymer, *Training School*, 17.
113. Ibid., 28–29.
114. Ibid., 30.
115. Ibid., 64–65.
116. Nightingale to Henry Bonham-Carter, 3 September 1865, cited in Baly, *Nursing Legacy*, 49.
117. Nightingale to Henry Bonham-Carter, 28 August 1875, *CW* 12, 305.
118. Elizabeth Torrance to Nightingale, 7 April 1872 & 12 April 1872, BL Add Mss 47749 ff249–252.
119. Torrance to Nightingale, 11 December 1872, BL Add Mss 47749 ff300; Mary Crossland to Nightingale, 15 September 1875, BL Add Mss 47738 f7.
120. Nightingale to Henry Bonham-Carter, 28 August 1875, *CW* 12, 305.
121. Seymer, *Training School*, 68; Mary Crossland to Nightingale, 10 August 1881, BL Add MS 47738 ff206–210.
122. Seymer, *Training School*, 86.
123. Nightingale's addresses to the Nightingale School probationers, 1874 & 1872; *CW* 12, 792, 766.
124. Nightingale's addresses to the probationers, 1874 & 1876, *CW* 12, 805, 828.

125. Nightingale's address to the probationers, 1900, *CW* 12, 880.
126. Vicinus, *Independent Women*, 87–101.
127. Nightingale, *Trained Nursing for the Sick Poor* (London: Cull, 1876), 2, 5.

CHAPTER 6

Home Front

On 28 March 1854, Britain and France joined the Ottoman Empire in its conflict with Russia and thereby began, in Orlando Figes's words, the first 'truly modern war'.[1] The Crimean War, as it became known after the Franco-British invasion of the Crimea in September 1854, was modern in terms of the nature of the combat. In contrast to the battles of movement of the Napoleonic Wars, this war was largely fought statically in trenches using rapid-fire weaponry and was marked by an absence of heroic figures. As the politician Sir George Sinclair noted, 'in the Cabinet, no Chatham; in the Navy, no Nelson; in the Army, no Wellington; in the Church, no Luther'.[2] The upper-class gentlemen of the army command—condemned by *The Times* as 'old men of the past'—oversaw the deaths of over 19,000 British servicemen, largely from avoidable diseases.[3] Perhaps the most recognisably modern aspect of this war, however, was its representation in the media. War 'has always been good for the news trade', but the Crimea coverage was far more detailed, vivid, and immediate than in previous conflicts.[4] Daily accounts in *The Times* enabled readers to not only read about the conflict from their homes, but actively contribute to it.

Recent scholarship has demonstrated the profound nineteenth-century cultural shifts of which this war coverage was a part. Stefanie Markovits has shown how new forms of written and visual representation brought a richer and more vivid experience of the war to the population at large.[5] Newspaper reports, poems, paintings, photographs, and even household objects elicited a new understanding of, and compassion for, the common

P. Crawford et al., *Florence Nightingale at Home*,
https://doi.org/10.1007/978-3-030-46534-6_6

soldier. The 'military men of feeling', to borrow Holly Furneaux's phrase, were brought into a far closer and more sympathetic relationship with the British public.[6] Less attention, however, has been paid to the way that the stirrings of home, intimacy, and compassion that characterised this mid-century culture of war enabled Florence Nightingale to undertake the work for which she became best known. The iconic image of Nightingale as the lady with the lamp emerged from this newly sentimentalised context and the gendered and class-based associations with the home that went along with it. But rather than limit the scope of Nightingale's action and influence, such emotive casting often enhanced it. Mary Poovey has argued that Nightingale demonstrated 'the almost limitless potential the domestic ideology contained to authorize aggressive projects that far exceeded the boundaries of the home and even of England itself'.[7] Discourses of home supported a practical and symbolic alliance that united Nightingale with the British state, monarchy, and, crucially, reading public in their varied attempts to reclaim control and accountability for the war from those 'old men of the past'.[8] During its early stages, Nightingale worked with the media to open the experience of war to a far wider portion of the population. Her popular mission was to carry the comforts of home, beside her household lamp, to the forgotten soldier.

In short, Nightingale worked within a wider discourse of home during a war that was consumed and, in some senses, commanded from the drawing rooms of England. She fostered both the sympathies and the practical enterprise of the virtual readership residing there. Her rise to iconic status was by no means inevitable, but instead emerged at a cultural moment, drawing on the diverse and at times contradictory meanings of home that were most alive in British society at the time of the Crimean War.

Enthusiasms at Home

The reasons for Britain's participation in the Crimean War—protecting trade routes and preventing possible Russian expansion into the Balkans and Bosphorus—were relatively distant from the concerns and lives of the wider British population. Something more was required to bring those reasons closer to home and create 'the feeling of the war being just', as the poet Arthur Hugh Clough, later Nightingale's secretarial assistant, put it in February 1854.[9] It was not necessarily a straightforward proposition to take up arms on behalf of the clearly declining, Muslim Ottoman Empire to fight against a fellow Christian nation in a mysterious place called 'The East'—the vague, orientalist term used to describe the theatre of conflict. Legitimacy for the conflict was established through the idea that the Russians threatened British, liberal values. Nightingale contributed to what became a far-reaching media campaign to convince the British population of the 'Russian menace' when she offered to Charles Dickens's magazine *Household Words* an account of Russian outrages that she had collected during her stay in Rome in 1848.[10] In May 1854, Dickens published an article based on her testimony. 'The True Story of the Nuns of Minsk' described the Russians subjecting an order of Catholic nuns who refused to convert to the Russian Orthodox Church to whippings and withholding food and clean water.[11] This piece helped cement the idea of the Russians as ruthless authoritarians, determined to suppress the legitimate desires of national and religious minorities. *Household Words* later printed a further article, 'At Home with the Russians', which assured its readers that any civilisation in the enemy's country was 'not more than skin-deep'.[12]

The commencement of military hostilities brought the war even closer to home. *The Times* was the most prominent of the newspapers to capitalise on Britain's confidence in its liberal convictions and to speak directly to the public in their homes. It represented itself as a defender of the common Briton against Russian-style abuses of power, allowing, as it printed in November 1855, the 'intelligent and self-respecting races' (i.e. its British readers) to freely scrutinise the decisions made on their behalf during the war—the government was slow to respond with the kinds of censorship that would be implemented in later conflicts.[13] These influential statements helped *The Times* to establish a contract of accountability with a significant portion of the public and to assume an unprecedented role in shaping public opinion.[14] By 1855, with the war well underway, it boasted a daily print run of over 60,000 copies, compared to the 5000 of its closest rival, *The Morning Chronicle*.[15]

A novel feature of this war was the influence of special correspondents, most famously William Howard Russell and Thomas Chenery of *The Times*. Their reports combined a range of testimonies and eye-witness accounts, which were sent by mail boat—not via telegraph, as is often supposed—to appear in print a 'mere' two weeks later.[16] As Markovits argues, the effect of this timely coverage was that it 'clearly struck those caught up in its events as "real time"'.[17] Wood engravings and lithographs appeared just as speedily: improved reproduction methods from the 1840s and a relaxation of the taxes on popular publications in 1855 enabled

Fig. 6.1 'Enthusiasm of Paterfamilias: On Reading the Report of the Grand Charge of British Cavalry on the 25th', John Leech, *Punch*, 25 November 1854, 213

drawings from artists such as Joseph Archer Crowe and Constantin Guys to become ubiquitous in *The Illustrated London News*, which by 1855 was attracting 200,000 readers per week.[18] While much is written about telegraphy and photography, it was these two forms, engravings and first-hand reports, that did the most to shape the public opinion of the war.[19] This montage of textual and visual representation became self-referential.[20] Famously, Russell's 25 October 1854 account of the Battle of Balaklava, the second of three bloody engagements in the first autumn of the war, reported how a 'noble six hundred' soldiers had fallen 'victim of some hideous blunder'—an interpretation immortalised in Alfred Lord Tennyson's tragic poem 'The Charge of the Light Brigade' a few weeks later.[21]

'Enthusiasm of Paterfamilias', a cartoon in *Punch* magazine (Fig. 6.1), shows the effect of this coverage upon readers at home.[22] A man in comfortable domestic surroundings is shown reading Russell's account of the 'noble six hundred' while his family huddle around the dining room table. The report provokes noticeably different effects, from the father wielding a poker, replicating a sword in battle; his sons waving their napkins in a show of patriotic support; to their mother sitting next to a toddler and crying. Trudi Tate has argued that this illustrates the '*fantasy* investment in war', whereby readers imagined the conflict in ways that transcended the routines and spaces of their regular domestic lives.[23] Nightingale acknowledged this phenomenon in a May 1855 letter describing her work in the war hospitals to her family at home, pausing to wonder 'what suggestions ... the above ideas make to you in [the] Embley drawing room'.[24] 'It is doubtful whether our soldiers would be maintained if there were not pacific people at home like to fancy themselves soldiers', wrote George Eliot in *The Mill on the Floss* (1860). 'War, like other dramatic spectacles, might possibly cease for want of a "public"'.[25] Enthusiasms like that of the poker-wielding paterfamilias prompted a wave of what Paul Hunter has called 'participatory journalism', whereby family members and soldiers commented on the war in letters published alongside the official newspaper coverage.[26]

The Times' war reportage undeniably provided readers with a form of entertainment, but the newspaper insisted that this was of a more serious variety than *Punch's* satirical cartoons, explaining that 'parents, wives, brothers, the whole family circle' had urged it to 'publish, and tell the whole truth'.[27] Once the conflict slowed to trench warfare, its reporting more than fulfilled this demand in its exhaustive accounts of everyday conditions that focused on the details behind the fighting.[28] Fanny Allen, a Nightingale family friend, wrote to Parthenope to explain that she could 'do nothing but read the papers now'. News stories 'brought things so vividly before our minds that we realise the miseries of war more', she explained, and instilled an 'intense feeling of interest in the suffering of the army that now appears so universal'.[29]

Nightingale responded in varied ways to the media's influential role in the war. She nearly always denied reading the papers herself, writing to her sister in June 1856 that she had 'not seen a paper for ten weeks' from 'sheer lack of time day or night'.[30] Yet anecdotal accounts suggest papers being spread around her Scutari room.[31] Moreover, *The Times* certainly influenced her decision to contribute to the war effort in the first place. In the 9 October 1854 issue, which Nightingale read at Lea Hurst, Russell turned his reporting away from 'the heroism' of the Franco-British victory at the Alma and towards his eye-witness account of 'the unhappy wounded, who have returned to die or to linger out their period of suffering within the hospitals'.[32] Nightingale wrote soon after: 'I do not mean to say that I believe *The Times* accounts, but I do believe that we may be of use to the wounded wretches'.[33] Even if her fascination was combined with doubt, her statement demonstrated the media's role in bringing the circumstances of the Crimea to her awareness. Indeed, it was this capacity to reach a wide readership that made her 'almost wish to be a newspaper writer myself', describing it as 'an inspiring thing to be writing, not to a party as the *Record*, the *Standard*, the *Daily News*, the *Guardian* do, but to the world as the *Times* does'.[34] Nightingale left for London on 10 October, the day after Russell's landmark account, commenting to a friend on 'the state of the Scutari Hospitals' and explaining that 'she should like to go to help'.[35]

'Practical Experience'

Nightingale's introduction into the public consciousness came courtesy of the close relationship that the press had developed with the homes of Britain. A piece titled 'Who is Miss Nightingale?', which first appeared in *The Examiner* on 28 October 1854 (before Nightingale had even reached the war zone), described her as 'young (about the age of our Queen), graceful, feminine, rich, and popular', holding 'a singularly gentle and persuasive influence over all with whom she comes in contact'. The piece constructed Nightingale as a philanthropic aristocratic lady, drawing on the conventional idea that such women's privileged relationship to domestic space acted to elevate the moral atmosphere of wider society. 'Her happiest place is at home, in the centre of a very large band of accomplished relatives, and in simplest obedience to her admiring parents'. *The Examiner*'s rhetorical challenge was to introduce Nightingale's ambitious, distant mission in a way that would not rupture familiar, and appealing, discourses of home. The piece explained that it was Nightingale's love for home that drove her to care for those outside of homes, firstly in a London institution for those suffering from 'homelessness' (Harley Street), and now in responding to a 'cry of distress' from further afield. By responding in this way, she would 'wider exercise' her 'sympathies, experience, and powers of command and control'.[36] Nightingale's early representation therefore repeated a familiar and clearly gendered trope of the brave yet angelic heroine whose allegiances were, first and foremost, with the home. The image, placing Nightingale within accepted expectations of womanly behaviour, could stir national pride, allowing her to venture into a foreign, violent conflict with her respectability intact. It also anticipated the role that Queen Victoria was to assume in the war coverage as a representative of maternal care.[37] Nonetheless, *The Examiner* acknowledged that some of its readers would find the idea of a lady travelling to the war zone 'eccentric or at best misplaced'. Lord Raglan, the overall commander of the army, expressed similar concerns, questioning if she would cope with 'painful scenes' and not miss the 'comforts' to which she was accustomed.[38] A letter to *The Times* signed 'common sense' similarly doubted if one 'born and bred in the enjoyment of every luxury' would have either the 'ability to endure fatigue' or the will to address 'official mismanagement and neglect'.[39]

The man responsible for officially inviting Nightingale to work for the army was Sidney Herbert, a cabinet minister in the War Office. In October 1854, Herbert was facing potentially devastating criticisms about the neglect of the nation's soldiers, which the newspapers blamed on the incompetence of the upper-class army command and the government's inattention. While the invitation to Nightingale presented him with the opportunity to address some of the deficiencies of food, clothing, shelter, and medical services for which he was responsible as Secretary at War, it had to be seen to emerge from the camp of common sense and practical reform, not the upper-class patronage networks that had brought about catastrophe in the first place. 'It was the business of Government to provide proper nurses for military hospitals', warned *Punch* magazine, 'not to leave the duties of the soldier's nurse to be undertaken by young ladies of rank and fashion, who knew not even as yet what it was to nurse a baby'.[40] While Herbert knew from his wife, Elizabeth, a member of the Ladies' Committee of the Harley Street Establishment for Gentlewomen During Illness (Chapter 5), that Nightingale possessed the practical skills and administrative ability required, at this early stage any public statement had to ensure that Nightingale symbolised the values of hard-headed domestic management, not those of aristocratic decadence.

Once Nightingale had agreed to participate and the initial arrangements for her departure were in place, Herbert described her in *The Morning Chronicle*, a paper that he owned, in terms of practical reform: as possessing 'greater practical experience of administration and treatment than any other lady in this country'.[41] Writing a few days later in *The Times*, Herbert underlined Nightingale's distance from 'ladies' with 'generous enthusiasm' who were 'little aware of the hardships they would have to encounter'. He emphasised that Nightingale was 'ready to yield that implicit obedience to orders so necessary to the subordination of a military hospital'; she was to be 'the one authority to select, to superintend, and direct in the British general hospitals in Turkey a staff of female nurses'.[42] Both articles played down Nightingale's aristocratic background in favour of the professional language of the public servant.

Nightingale selected the thirty-eight nurses that travelled under her authority based on the 'practical experience' that Herbert had emphasised, with roughly equal numbers of nurses from Anglican sisterhoods, Catholic nunneries, and secular, civilian hospitals.[43] In one area, however, Nightingale and Herbert disagreed, namely the scale to which her operation could be extended. Herbert believed the rapidly increasing number of injuries and illnesses demanded more nurses. But Nightingale had always been wary of overseeing a large team and, in a letter to Selina Bracebridge, her early confidante who now supported her in her Crimean mission, had stated that it would be 'infinitely easier to pioneer the way with three or four women than to march in … with a great batch of undisciplined women, not knowing what places to assign them'.[44] While she was willing to increase this figure once she became aware of Herbert's ambition, the thirty-eight nurses that she was eventually tasked with overseeing was, as she saw it, an attempt at 'governing that which cannot be governed'. 'Half that number would be more efficient and less trouble', she explained.[45] Nightingale probably feared that, the larger the operation became, and the further removed from her close oversight, the greater the likelihood that some serious mistake or scandal would arise which might imperil the whole cause of trained female nursing. Once a second group of forty-six nurses led by Mary Stanley was sent from London Bridge on 2 December 1854, Nightingale was forced to work within an even larger structure over which she struggled to assert overarching control. This moved her beyond the domestic models of smaller institutions and homes to the dizzying scale of a complex healthcare institution. The analogy between the management of a household and that of the army's nursing force was under strain.

Nightingale eventually adapted to this larger workforce by following a principle of 'unity of action & personal responsibility' that she expressed in a letter to Herbert on 25 November 1854.[46] 'Unity of action' meant assigning a central figure of oversight, or, as she explained two months later, '*a head* – some*one* with *authority* to mash up the departments into uniform and rapid action'.[47] This wish to direct and exercise control was characteristic of Nightingale's approach. She believed that the leading figure in the army hospitals had to be able to adjust processes according to the particular circumstances on the ground, rather than from England, and for this reason wrote that 'the grand administrative evil emanates from home', denying the army a 'central authority capable of supervising and compelling combined effort for each object at each particular time'.[48] Home was instead to be brought to where the action was. In early January 1855, Nightingale outlined how this new operational structure was to be realised. Initially, she argued that the ideal purveyor should be like 'paterfamilias',[49] but later that month she divided the role into three subsidiary functions that had some echoes of the large English country house: one was 'to provide us with food'; another held responsibility for 'hospital furniture and clothing'; and the last was to be a 'hotelkeeper, a house steward, who shall take the daily routine in charge'.[50] Nightingale hoped to transform the operational structure of the military by extending and expanding household structures and analogies with which she was familiar.

'A Few Creature Comforts'

However, upon arriving in early November 1854 at the Barrack Hospital in Scutari, Turkey (today Üsküdar, a district of Istanbul), Nightingale faced a more immediate crisis. As thousands of casualties arrived by ship, having survived a harrowing journey of hundreds of miles across the Black Sea from the Crimea with little to no medical attention on board, she soon found herself short of supplies. 'Not a sponge, nor a rag of linen, not anything have I left', she exclaimed, continuing, 'we have not a basin nor a towel nor a bit of soap nor a broom'.[51] The shortage was exacerbated by the catastrophic sinking on 14 November of the *Prince*, a ship carrying the army's provisions of winter clothing and medicines. Sydney Godolphin Osborne, who conducted an unofficial inspection of the military hospitals, described finding a 'vast field of suffering and misery' that lacked 'the commonest provision'.[52] Osborne observed soldiers being given raw meat rations without means to cook them, while Nightingale later testified that soldiers were forced to 'tear their food like animals' due to 'the deficiency of knives and forks'.[53] Nightingale ordered her nurses to make flannel shirts and bandages while herself assuming the position of, as she wrote in January 1855, 'a kind of general dealer in socks, shirts, knives and forks, wooden spoons, tin baths, tables and forms, cabbage and carrots, operating tables, towels and soap'.[54] But this did not mitigate the more structural problems of supply for which the Purveyor-General, Matthew Wreford, was responsible.

'There is, no doubt plenty of everything, but nothing to be had', reported the Bracebridges to the Nightingale family on 10 November 1854.[55] The problem was that the Purveying Department was not designed to prioritise the demands of the increasingly desperate soldiers. Being under the ultimate control of the Treasury, the department was instead driven by what Nightingale called a 'deity of cheapness' that made compliance with process and economy the guiding principle: as Nightingale put it, 'the correctness of their bookkeeping [w]as the primary object of life'.[56] The consequences of their resulting unwillingness to release supplies were serious. The cold winter weather threatened soldiers who could not access warm clothing and coats that remained unpacked.[57] 'I have no compassion for the men who would rather see hundreds of lives lost than waive one scruple of the official conscience' wrote Nightingale.[58]

It was some months before Herbert was able to bring the Purveyor under his command in the War Office. In the meantime, he followed Nightingale's suggestions and prompts by advocating for more flexibility and enterprise. 'This is not a moment for sticking at forms', Herbert wrote to the Purveyor-General, 'but for facilitating the rapid and easy transaction of business'.[59] *The Times* was already preparing the public for a battle against the 'red-tapeism' of an out of touch army command, in December 1854 listing features of this 'grossest mismanagement of the war': 'incompetency, lethargy, aristocratic hauteur, official indifference, favour, routine, perverseness, and stupidity reign, revel and riot in the camp before Sebastopol, in the harbour of Balaklava, in the hospitals of Scutari, and how much nearer home we do not venture to say'.[60]

In its reports, *The Times* related this 'grossest mismanagement' to two discourses of home which had become a feature of its war coverage. First, home could bring comfort and restore a sense of compassion to the stranded and neglected soldier. Second, home provided an ethic of management, powerfully framed as part of a national effort, that could improve the dire situation of the British army in the Crimea. For these propositions to make sense, the paper had to develop a close and sustained connection between the hardships that the soldiers were experiencing and the more comfortable settings in which the paper was read. 'Here we are sitting by our firesides, devouring the morning paper in luxurious solicitude' began the lead column of 12 October 1854, 'counting the days of Sebastopol'. 'To us war is a spectacle', it continued, identifying a troubling distance between the reader and the frank accounts of 'the thousands of poor wretches who have to pass the night of victory or defeat with no friendly hand to bind their wounds or slake their deadly thirst'. The lack of care shown to the soldier was then made an explicit problem of home—a domestic threat—such that 'the almost total disorganisation of our army in the Crimea, and its awful jeopardy' was said to come 'not from the Russians, but from an enemy nearer home – its own utter mismanagement'. Having enabled the reader to imagine and feel the suffering of the army soldier, and then constructed the resultant crisis as a failure of household management, *The Times* announced a means for the public to participate in the conflict—an 'opportunity' for 'showing how your hearts beat for your fellow-countrymen and your noble allies' by 'sending them a few creature comforts'.[61] In her analysis of this 12 October issue of *The Times*, Stefanie Markovits observed how the pronoun 'we'—as in, 'we are sitting by our firesides'—changed to the more direct 'you' at just the moment that the reader is called to action.[62] This writing cleverly fashioned the reader at home as a participant in, not merely an observer of, the war.

The public response to this appeal was immediate. The next day's issue of *The Times* included offers of support ranging from hundreds of pounds in donations to '£5 in money and some old sheets'.[63] Such donations were quickly organised under a more formal public appeal that became known as *The Times* Sick and Wounded Fund, with John C. MacDonald, a *Times* correspondent, appointed as almoner. This was not the first appeal of the Crimean War—the Royal Commission of the Patriotic Fund had already been established to support war widows and their children—but it was the one most clearly driven by popular feeling. It also represented a distinct shift in the role of the press. In 1863, some years after, *The London Review* noted that 'the most convenient organ of publicity' had for the first time been 'converted into the most powerful financial agent for performing certain laudable works'.[64]

MacDonald soon had many thousands of pounds to distribute as he saw fit. Thanks to a family connection between his colleague Henry Reeve and the Nightingales, he had been able to travel to Turkey with the first wave of nurses. The Sick and Wounded Fund then relied on Nightingale to direct resources towards those supplies that would be most valued by the soldiers. With this support, she was able to establish an independent storeroom in her own quarters and send MacDonald to buy supplies in local markets.[65] By 30 November 1854, Nightingale had obtained '2000 shirts, 250 pairs of sheets, 400 flannel vests, 10 pieces of flannel, 365 warm quilted coats, 72 worsted jackets, 1200 pairs of stockings, and 400 towels'.[66] Along with other private donations, the fund enabled a quick, direct line of supply. No doubt sensing the symbolic as well as practical effects of this arrangement, Nightingale appealed 'to the public' in her own name in the 25 November issue of *The Times* and explained that 'lint and old linen, or linen rags, are still urgently required', hoping that 'we shall not appeal in vain to our readers for a large supply of this most necessary material in the treatment of wounds' (Fig. 6.2).[67]

Nightingale worked in close co-operation with *The Times* throughout this first winter of the war. Their alliance drew on a middle-class sense of responsibility that connected the public and soldiers with high-profile volunteers, not only Nightingale but figures such as the London-based chef Alexis Soyer who mobilised similar household enthusiasms with his

Fig. 6.2 Wood engraving showing reception of lint and other items for the wounded at Scutari, *The Illustrated London News*, 30 December 1854, 701

'Culinary Campaign' that, he claimed, showed an 'energy and perseverance required to eradicate old and obsolete systems'.[68] Through her pursuit of domestic standards, Nightingale embodied the wishes of 'relatives and friends of the sufferers' and, as *The Times* wrote just after the war had ended, effectively came to represent 'public opinion' itself.[69] The practical aspects of the collaboration halted around the spring of 1855, however, when Nightingale expressed frustration that the newspaper was overstating its role as the almoner to the army. '*The Times* are playing a most unfortunate game', she wrote to Herbert on 5 February 1855, 'I am told it is always writing to prove that it has done everything', when, she explained 'not above one half of the things supplied by me comes from the "Times" fund'.[70] Nightingale also resented being 'dragged before the public' in her now quite frequent appearances in the paper.[71]

The Times Fund characterised the 'middle-class assertiveness', which, as Olive Anderson has argued, was a vital feature and effect of the war.[72] This assertiveness was felt in Parliament, when the radical MP John Arthur Roebuck made the question of supplies central to a motion of 29 January 1855 that led to the resignation of the Prime Minister, the Earl of Aberdeen. Organisations such as the Administrative Reform Association continued in the spirit of Roebuck's motion, calling for public affairs to follow 'those principles of common sense which the practical experience of private affairs suggests as essential to avert failure and ensure success'.[73] Several commissions, notably that looking into the army's supply situation led by Sir John McNeill and Colonel Tulloch, sought to bring similar improvement and accountability to the conduct of the army at war. These commissioners arrived in the war zone on 6 March 1855 to investigate 'the mode by which supplies of food, forage, and any other articles are obtained', and produced a report that Nightingale judged to be 'correct, cool, dispassionate'.[74]

As the state assumed responsibility for improving the supply of core resources to the army, the public were asked to make increasingly sentimentalised and symbolic contributions. The involvement of Queen Victoria, who saw in the popular campaign an opportunity to exercise one of her 'dearest prerogatives', was crucial in this regard.[75] She had read about the plight of the soldiers and sympathised with the newspaper accounts that evoked a sense of pity and compassion for them. This was an opportunity for her to publicly display her care and concern to the ordinary soldier in a comparable manner to Nightingale herself. On 6 December 1854, the Queen wrote to Herbert to express her wish that 'Miss Nightingale and the ladies would tell these poor, noble wounded and sick men that *no one* takes a warmer interest or feels *more* for their suffering or admires their courage and heroism *more* than their Queen'.[76] This statement was printed in *The Morning Post*, and, per Victoria's express wishes, copies were posted onto every ward in Scutari, to an apparently rapturous reception from the soldiers.[77] Soon after, Nightingale received correspondence from the Keeper of the Privy Purse enclosing 'some packages containing some comforts and useful articles which Her Majesty wishes to be placed in your hands for distribution'. While Nightingale's mission to provide the practical essentials had been 'observed by the Queen with sentiments of the highest approval and admiration', the enclosed objects—books, newspapers, periodicals, air cushions, and woollen blankets—were notably more luxurious. The Queen had directed these articles to Nightingale, the Keeper explained, because, as they 'did not come within the description of Medical or Government stores, usually furnished, they could not be better entrusted than to one who, by constant personal observation, would form a correct judgment where they would be most usefully employed'.[78] Unlike the basic supplies like bandages and shirts obtained by Nightingale and MacDonald in the markets of Constantinople, the Queen's parcels represented comfort. As such, they presented a subtle but clear suggestion that popular campaigns, at least those with which monarchy were involved, were henceforth to complement, rather than replace, the official lines of supply. Nightingale appeared to accept this symbolic shift a few weeks later when she suggested that some wool provided by her be used to produce comforters on the soldier's beds, as 'something which the man will feel as a daily extra comfort which he would not have had without her'.[79]

Lord Ellesmere's Crimean Army Fund, set up in time for Christmas 1854, provided the nation with the means to comfort their soldiers in similar ways. Despite its name, Ellesmere's Fund called for material rather than financial donations. At this festive time, the trappings of home were a stark contrast to the reality of those soldiers 'plunged into the inevitable miseries of a winter campaign' while, wrote *The Times*, 'not a soul seems to care for their comfort'.[80] Like many charitable efforts, the benefits were felt by the giver as well as the recipient. 'We are all better with the desire to do something for the poor soldiers in the Crimea', wrote Lady Charlotte Bridgeman on 2 December, explaining that she had 'frantically begun knitting muffetees and comforters'.[81] While these calls for supplies tended to originate from London-based papers and politicians, the appeal mobilised local communities across Britain. 'There is scarcely a parish or hamlet of this county that has not furnished some portion of the army in this unfortunate campaign', *The Times* reported, also printing the advice from a 'Yorkshire Woman' to the 'Women of England' on the production of 'Mitts and Socks for the Army in the East', which advised that 'the best division of labour is for the mitts to be done by ladies, while (the yarn being furnished) the poorer women of the town or village do the socks'.[82]

Nightingale's sister, Parthenope, met a 'flood' of what she called 'English benevolence' when spontaneous contributions from family, friends, and strangers arrived at the Nightingale family home during and following Ellesmere's appeal.[83] Ginger brandy was sent from a distiller in Borough Market, London; knitwear including 150 waistcoats and 100 nightcaps came from the 'English Ladies at Pau'; and a box containing raspberry preserves and ginger biscuits arrived from a Mrs Gallop of Beaminster, Dorset.[84] A record of 'Contributions from Derby and Derbyshire', still held among the family correspondence, listed contributions from the region's leading industrial families such as the Arkwrights and the Strutts. Pocket handkerchiefs, tobacco, pipes, shirts, stockings, mitts, and muffatees were collected and sent in three boxes to the Crimea.[85] Parthenope's account suggests how the donations changed throughout the year: 'in December it was preserved meats, wine and money. Now we have reverted to shirts, knitted things and books'. They came from 'every kind and sort of person', she wrote, so that 'the sugar refiner's clerk and his 6 shirts' contrasted the 'small clergyman's wife's packet of little books and stockings'.[86] Parthenope occasionally expressed frustration at the sheer volume of material that she was asked to administer and direct, writing to her friend Ellen Tollett after many months of this task that a 'second blast of linen and knitted socks was nearly the death of me'.[87] Yet the letters also suggest Parthenope gaining a sense of purpose from this opportunity to participate from home in her sister's great deeds and feeling at least partial relief from the tensions that had existed when they lived at close quarters (Chapter 3).

Ellesmere's yacht, the Erminia, reached the Crimea on 13 February 1855 and was followed by two subsequent steamer ships within fifteen days that in total carried 1000 tons of goods on behalf of the nation for the 'comforting of their troops'.[88] The actual use of these contributions from 'our fair sympathizers at home', as Russell at *The Times* called them, seems to have been limited.[89] The coldest weather had ceased by the time of their arrival and one soldier described being left with 'enormous quantities of useless clothing ... fit only for a polar expedition'.[90] It was instead the symbolic value of 'trunks, packed up in village halls' being 'bundled into the huts of the men' that was important. Frederick Robinson, an officer in service at the time, described the almost miraculous arrival of these homely items:

> Out of the tissue-paper and sawdust came pots of honey, peppermint lozenges, arrowroot and ginger, messages of love and encouragement carefully written on scented paper, improbable-looking combinations and bits of knits, handsome hassocks and velvet smoking-caps, knitted waistcoats, and woollen helmets which muffled the ears and filled the mouth with fluff.[91]

The luxury items roused a nostalgia for Britain that was much like the 'longing' that Nightingale sometimes felt at Scutari for the 'western breezes of my hilltop home'.[92] Receiving packages lessened the psychological distance that the soldiers felt from Britain. Much of the clothing had been produced by soldiers' families and, as Furneaux writes, can therefore be thought of as 'creative collaborative practices' that enabled soldiers and their families to 'maintain shared cultures and a form of togetherness at a distance'.[93] The wide-ranging campaign for supplies that galvanised the British public showed how a large and complex issue could be addressed with a response that reassured each family, each householder of Britain, of the common sense and the compassion contained in their home.

The Lady with the Lamp

The image of Nightingale that took hold in the popular imagination followed on from this role as the co-ordinator of homely comforts. Its origins lay in a piece by MacDonald in *The Times* on 8 February 1855. The public uproar at the procurement failures that had dominated the autumn and early winter coverage had by this time turned to outrage at the number of soldiers dying from disease. Nightingale was therefore not only required to overcome delay and inefficiency, but also to be seen to ward off the threat of spreading contagions. MacDonald's article described Nightingale as a 'benignant presence' and 'influence for good comfort even amid the struggles of expiring nature'. She was a 'ministering angel', he continued,

> and as her slender form glides quietly along each corridor every poor fellow's face softens with gratitude at the sight of her. When all the medical officers have retired for the night, and silence and darkness have settled down upon those miles of prostrate sick, she may be observed alone, with a little lamp in her hand, making her solitary rounds.[94]

This extract presented Nightingale as a figure of grace and spiritual comfort in ways that seemed to extract her from the realm of war entirely. The soldiers lied 'prostrate', unable to rise from their beds due to illness, while Nightingale glided effortlessly, 'quietly along'. Under the cover of darkness, her mere presence was enough to bring relief to the faces of the ill as they beheld her.

The night-time setting became more pronounced two weeks later in an engraving in *The Illustrated London News* (Fig. 6.3), inspired by the description above.[95] Light is crucial to this iconic image, borrowing emphasis from the monochrome photographs that were then emerging from the war, as well as from earlier traditions of realist painting. 'The night scenes are quite Rembrandt' wrote Selina Bracebridge from one of the Scutari hospitals the previous autumn, anticipating a comparison between Nightingale 'with candles surrounding a poor fellow on the ground with his arm off' and the illuminated figures of the seventeenth-century Dutch painter's work.[96] In the 1855 engraving, Nightingale's face shines out over an injured patient, whose bright clothes and bedsheets accentuate the intimate exchange between nurse and soldier. By the time the image appeared, some of the personal letters of condolence that Nightingale had written to the families of deceased soldiers had been printed in newspapers, helping to establish her public image as a conduit of family sympathies and guardian of the soldiers' welfare.[97] Yet the blackness that lies behind the illuminated bodies in the image suggests a scale many times larger than what could be easily conceived. Lying out of sight were the 'miles of prostrate sick' that MacDonald had described—no exaggeration of the four miles of beds that ran across the Barrack Hospital and the smaller General Hospital at Scutari.[98] Much like the 'wider exercise' of the home in *The Examiner*, the engraving imagined Nightingale spreading the small, intimate, and sacred during her 'solitary rounds', thereby addressing a problem that was far greater in scale.

Fig. 6.3 Wood engraving of Nightingale holding her iconic lamp, standing between rows of the sick and injured in a hospital at Scutari, *The Illustrated London News*, 24 February 1855, 176

Another important effect of the 'silence and darkness', described in *The Times* and reimagined in *The Illustrated London News*, was the attention that it drew to a miniature emblem that, more than anything else, came to be associated with Nightingale: 'a little lamp'. The lamp has tended to be understood in terms of its religious overtones, yet it also held more everyday associations.[99] It was an object that featured in readers' daily lives; a symbol of English family life that Mai Smith thought to describe to her children as 'the "little lamps"' during her time accompanying her niece at Scutari.[100] At the same time, Mark Bostridge has pointed out that Nightingale's lamp was made up of characteristically Islamic features and styles unfamiliar to the English reader.[101] Indeed, what made the lamp such an effective symbol in early 1855 was its ability to bring the worlds of the home and the faraway war together, thereby demonstrating that, as the literary critic John Plotz has argued in his work on portable property in the Victorian period, 'infallible, unbreakable relics might operate as a moveable repository of both family feeling and Englishness'.[102]

It was, of course, a fiction that Nightingale could meaningfully tend to so many bodies. She was just one individual circulating on 'solitary rounds' among the many sickbeds.[103] But the image was nonetheless effective in providing a prism through which the public could imaginatively relate to the nation's soldiers, addressing the inevitable gap in understanding between the British population and the realities of life in the army camps and hospitals. War was, of course, unlike a drawing room; it was a mission into foreign land, characterised by the unknown, and typically driven by masculinised notions of bravery amidst hardship.[104] '[W]hat had women to do with war?' asked Lytton Strachey of Nightingale in *Eminent Victorians*.[105] It was the ideal of home that helped to keep these seemingly contradictory ideas together. Frederick Edge's *Tribute to Florence Nightingale* (1864), one of the many eulogies to appear after the war, presented a fictional soldier repeating the objection that 'you cannot have the comforts and the care of "Home" on the grim fields of war'. To this, Nightingale was imagined as responding, 'Where I am is "Home"; I bring with me its comforts and its care to the battlefield and camp, and all a mother's love shall tend your aching brow and staunch the oozing blood'.[106] Nightingale's sourcing and provision of comforts became the most distinctive and popular aspect of her work in the war.

Homecoming

By the middle of 1855, the British army and its hospitals had been able to incorporate at least some of the order and comfort of home. The worst of the war and scandals about supplies and sanitation had passed. A significant proportion of the home population had either donated money or homely items to help improve the previously grim conditions. The public mind now began to wonder how it would understand and draw lessons from the war.

In summer 1855, the Herberts suggested that a 'testimonial of a substantial character' be created to honour Nightingale's 'noble exertions'.[107] After some discussion, it was decided that this would take the form of a public subscription to be spent on a training school for nurses that Nightingale would then oversee upon her homecoming: the Nightingale Fund (Chapter 5). The appeal spread to every corner of the nation, with 20,000 circulars distributed in the first month and the bookseller W. H. Smith placing 1000 leaflets promoting the appeal on his railway station bookstalls. It extended the kinds of homely enthusiasm that had galvanised householders across the country in the war appeals of the previous years. As the politician Lord Stanley stated in a speech in its honour, 'there is no part of England, no city or country, scarcely a considerable village, where some cottage household has not been comforted amidst its mourning for the loss of one who had fallen in the war'.[108] When the discussion turned to how the fund was to be spent, the *Derby Mercury*, which had previously largely mirrored the London papers in its coverage of Nightingale as a symbol of the nation, printed 'one or two words … with regard to our own county':

> To the country at large the object for which those subscriptions will be asked is a truly national object. To Derbyshire it is this and, also, something more. FLORENCE NIGHTINGALE is a native of Derbyshire. Hers was a familiar name among us, long before it had become a name familiar to and revered by England, and almost worshipped by thousands of England's noblest sons. And now that the name of FLORENCE NIGHTINGALE has attained an historic eminence which will for ages shed a lustre on her native county, surely Derbyshire should be among the foremost ranks of those who do her honour.[109]

Nightingale was presented as, at once, a national and a regional figure. The paper argued that, as well as London institutions, Derbyshire should receive at least some benefit from the funds raised. Such often-neglected regional accounts show how Nightingale commanded responses from distinct but overlapping communities—both the nation as a whole and what the *Derby Mercury* in the same article called 'the sympathy of a particular county'.[110]

Many of the images and words of 1856 reflected how the effects of homecoming were felt in the region and the home. John Everett Millais's painting *Peace Concluded* (Fig. 6.4), shows an injured soldier reading the newspaper while his young child plays with a set of Noah's Ark toys; each animal representing one of the major nations in the conflict.[111] Numerous children's toys and books responded to the war, including a series of images titled 'Incidents of War' and the *Panoramic Alphabet of Peace* (1856, Fig. 6.5) that represented Nightingale under the letter 'N' as the one 'who solaced our sick in a far distant land'.[112] The war's appearance in the culture of childhood helped to transfer the events of the front into

Fig. 6.4 *Peace Concluded*, an oil painting depicting a soldier from the Crimean War holding a newspaper, while his children play with miniature toys. John Everett Millais, 1856, Minneapolis Institute of Arts, Putnam Dana McMillan Fund, 69.48

Fig. 6.5 Image of Nightingale as part of a commemorative book for children marking the end of the Crimean War and the resumption of peace. *The Panoramic Alphabet of Peace* (London: Darton, 1856)

the intimate and smaller setting of the home at a time when the nation's families were adapting to the peace.

Nightingale was by this time established as the 'the heroine of the cottage, the workshop, and the alleys' who stood at the centre of this consumption and commemoration of war.[113] There was a significant market for Crimean ornaments and trinkets to decorate family homes, reversing the direction of objects such as slippers and sheets that had previously been sent to the front. The eventual capture of Sebastopol in September 1855 had brought a flurry of artefacts. Soldiers plundered everything from crockery to chandeliers; 'every kind of domestic utensil', as one officer put it, was looted as 'prizes much valued'.[114] Nightingale's image appeared on a dizzying range of media and objects that emerged from the later stages of the war onwards. She was represented on approximately 250,000 *cartes-de-visite*—small photographs around nine by six centimetres pasted onto cardboard.[115] Cheap portraits proliferated; a 'Nightingale cradle' for babies was available; and three different mass-manufactured porcelain figurines in 1855 also catered for an eager public.[116] These 'Staffordshires' show Nightingale in a range of poses, from comforting a wounded soldier to holding two cups on a tray. Such ceramics depicted public events and figures in handheld forms that could be displayed in the homes of everyone but the poorest, thereby bridging, as Rohan McWilliam has suggested, a 'twilight zone between the public and private'.[117] Such objects were produced *en masse*, without care for accurate portrayal and with no acknowledgement of the ownership of the original image—a fact that frustrated Nightingale's relations.[118] Nonetheless, it was through these objects of popular culture that Nightingale's image entered the everyday lives and homes of the British population.

If this flurry of consumption cheered some family homes, others were affected by a more sombre mood. A painting by James Collison titled *Home Again* (1856) shows one weary soldier's return from war to a crowded cottage scene. His family expresses worry and concern at their prospects, for, as the caption of the painting explained, his blindness would bring hardship on them all. A reflective mood affected the national level, too, once the peace treaty was signed in Paris on 30 March 1856. Beyond restricting Russia's naval presence in the Black Sea, it was not clear what had been gained. One in five of the 98,000 British soldiers who left for the Crimea had not returned home.[119] Queen Victoria admitted that 'peace rather sticks in my throat'.[120] To salvage some good feeling from the victory, the authorities announced a 'homecoming season'.[121] It ran from March to July 1856 and emphasised the theme of peace above nationalistic jubilation.

A public holiday on 29 May 1856 combined with the Queen's birthday to mark the midpoint of these homecoming celebrations. In London, events were organised with new technologies that included the illumination of public buildings and firework displays that showered 1000 rockets over Primrose Hill.[122] But papers such as the *Derby Mercury* reported parallel 'Peace Rejoicings' that were held in smaller, regional communities on the same day and have been far less studied. Derbyshire towns and villages from Ashbourne to Egginton witnessed everything from church bells, army bands, and feasts to the flying of the Union Jack and performances of the national anthem. Festivities in and around the Nightingale land in the Derwent Valley were marked by more personal celebrations acknowledging, as one chairman of an event in this area put it, 'a neighbour of our own, of whom we should be proud'.[123] The 'celebration of the restoration of peace' organised in Matlock Bath and Cromford, marked by a mile-long procession attracting approximately 5000 people, was the largest. It included a

> somewhat cleverly dressed effigy, intended to represent her whom the inhabitants of this neighbourhood peculiarly delight to honour – Miss Nightingale – which was carried through the village in the evening, and may be considered as a rude but hearty mode of expressing the admiration so warmly felt by the people generally for that excellent lady.[124]

A Nightingale effigy, garlanded with flowers spelling 'The Good Samaritan', was also noted by Julia Smith, her aunt, in a letter to Nightingale's father describing a celebration on the family's estate for the villagers of Dethick, Lea, and Holloway. This event had attracted some 800 local residents, from miners to farmers, and a sketch that Smith included in her letter showed musical bands, a long banner inscribed with the word 'Peace', and large marquee tents pitched on the hillside (Fig. 6.6). The

Fig. 6.6 Sketch of the 1856 peace celebrations in Lea/Holloway, Derbyshire. From a letter by Julia Smith to William Nightingale, Claydon House Trust, Nightingale Collection 210

event ended with cheers of 'three times three for Florence Nightingale' and Smith anticipated that the day would 'answer some fine purposes and be remembered for a lifetime'.[125]

Nightingale retained her standing as a member of the landowning family and was still understood as a familiar neighbour in the Derbyshire area. But these longstanding social networks also engaged with the images of Nightingale addressed to the nation. For instance, the textile manufacturer John Smedley, whose mill stood on Nightingale land, distributed her image locally, with 'every farmhouse and cottage' on the family's estate at Pleasley also receiving a print.[126] Spencer Timothy Hall introduced the 'heroic Maiden of Lea Hurst' in a *London Journal* article titled 'The Home of Florence Nightingale' (1855), and books such as Llewelyn Jewitt's *Stroll to Lea Hurst* (1855) and James Croston's *A Pilgrimage to the Home of Florence Nightingale* (1862) soon followed, bolstering the appeal of Nightingale's Derbyshire background among a wider readership.[127] In 1856 a letterhead showcasing Lea Hurst as 'The Home of Miss Nightingale' was, according to its creator, the Derby photographer Richard Keene, met 'with a ready sale' (Fig. 6.7).[128] The pressure that these larger forces and interests exerted on the intimacy of the Nightingale home were pronounced a few months later when 'mythical Florence', as Parthenope once described her, chose Lea Hurst as the destination for her own homecoming from the war.[129]

Over the summer of 1856, there was intense speculation about the whereabouts of 'the angel of mercy', who, as Lord Ellesmere reminded Parliament at the time, 'still lingers on the scene of her labours'.[130] 'Everybody worries, worries, worries about our coming home' Nightingale wrote a week after, clearly frustrated by the popular interest while overseeing the departure of the remaining nurses and soldiers from Scutari.[131] On 3 June, the Secretary of State formally thanked her for her 'human and generous exertions' and offered to facilitate her return transport. Nightingale declined and wrote to her family the following month to explain that she was still unable to 'fix the day for coming home' and reminding them of her wish 'to get home quietly, without any body knowing it'.[132] Numerous letters among the extended family repeated Nightingale's wish to bypass any public fanfare. Her aunt Mai Smith, who spent several months at Scutari with Nightingale before accompanying her on her journey home, wrote of the 'dread of the receptions with which she is threatened'.[133]

Fig. 6.7 Richard Keene's letterhead (1856) displaying Lea Hurst, Derbyshire, as 'The Home of Miss Nightingale', Claydon House Trust, Nightingale Collection 246/2

In late July 1856 Nightingale finally left Constantinople on a French vessel, the *Danube*, on a week-long journey to Marseilles, using a pseudonym to avoid detection. After travelling through France, over the channel, and staying overnight at the Bermondsey convent of the Irish Sisters of Mercy in London that had supplied several nurses to the war effort, she arrived at Whatstandwell Bridge Station, less than two miles from Lea Hurst, on 7 August 1856. Edward Cook, Nightingale's first biographer, introduced the quasi-pastoral image of Nightingale walking 'up from the little country station' for the last part of her journey to Lea Hurst, which local historians have shown was much more likely to have been undergone by carriage.[134] Either way, the return was 'noiseless', as Nightingale's sister explained, 'surely disappointing the Derbyshire folk who had set their hearts on triumphant processions'. The only 'innocent greeting' came from 'a little tinkle of the small church bell' the next day at the Methodist Chapel at Lea.[135] Nightingale's plan was to avoid any coverage, yet this modest story soon inevitably attracted the attention of the national press. *The Morning Chronicle* was the first to report that 'after an absence of nearly two years from her native country' she had 'returned to her home in Derbyshire'.[136] Two days later, *The Times* printed similar detail and used the occasion to recall its depiction of Nightingale and her nurses as 'the representatives of home – of mothers, wives, sisters, and of all the sacred humanities of life'.[137] Homecoming gifts arrived at Lea Hurst from various groups of well-wishers, from the female tenants of the Nightingale estate at Pleasley to artisan cutlery-makers from Sheffield.[138] Nightingale drafted a standard response to these various admirers: 'your welcome home, your sympathy with what has been passing while I have been absent, have touched me more than I can tell in words'.[139]

For William Nightingale the public exposure of his daughter, herself averse to 'effigy and praises', made him 'tremble for my own name'.[140] He felt this aversion to public attention more deeply at Lea Hurst, for this had always been what he once described as his 'solace' from the world.[141] While there had been occasional visitors during Nightingale's time at Scutari, such as a veteran found lurking in Lea Woods to glimpse 'the home of Miss Nightingale', many in the family expressed concern that this attention would dramatically increase.[142] Hilary Bonham-Carter suggested installing 'No Entrance' signs to deter 'impertinent swarms' and recommended installing a telescope on a nearby hilltop for fans to inspect the family home from a distance.[143] Selina Bracebridge's idea was to 'put a stop to the peeping-in public' by stationing a 'policemen at the gate when she comes home'.[144] A more effective deterrent came thanks to the co-operation of the *Derby Mercury*, which, two weeks after announcing Nightingale's homecoming, asked 'inconsiderate persons' intent on giving her 'a public reception' by 'shouts and clapping of hands' to desist, her character being 'the very last to take delight in noisy plaudits'. She 'needs a long and complete rest. Let her have it. Let no one intrude upon her, with praise or projects', wrote the paper.[145] Nightingale, for her part, was beginning a period of enquiry and critical reflection that sharply contrasted the public enthusiasm that surrounded her home.

Indeed, Nightingale hoped to withdraw from public attention now that the war had ended. 'I desire privacy for the reason that I consider publicity to have injured what is nearest my heart', she wrote soon after her return.[146] The war had, she wrote, been 'acted out on the most colossal *scale* stage which perhaps it has been ever given to any nation or age to witness'. Nightingale's representation and involvement in the national media had successfully embraced this scale, extending the emotional and ideological reach of the home far beyond its typical dimensions to expose what she called 'the great Crimean catastrophe'. Yet the press campaigns were not enough to change a 'state of feeling' which, she continued, 'did exist, exists and will continue to exist at home'; they had not prevented her from coming home beside 'the remains of that lost army' to find the 'malefactors', as she called them, 'in all the official posts, honours and drawing rooms of the kingdom'.[147] Despite the tone of despair in these writings of 1856, Nightingale remained committed to honouring the lives of the soldiers who had died by bringing about change by other means. Her immediate focus was on army reform. Her monumental, 853-page response to the war, *Notes on Matters Affecting the Health, Efficiency and Hospital Administration of the British Army* (1858), barely known today, contained many of the suggestions that made their way into the other commissions of the period. It attempted to translate the popular disgust at 'the state of the army [that] has been constantly before the public' into a series of probing questions that the often-sensationalised coverage had not addressed: 'were the rules and regulations of the service adequate to their object, the preservation in health and life of the British Army?', Nightingale asked; 'What are the rules and principles now? What will they be for the future?'[148] Such questions were inspired by Nightingale's conviction that some insight, some improvement, could emerge from the terrible suffering of the war. The quiet, enclosed parameters of the home were where her work towards this objective would commence.

The home had provided Nightingale with a motif with which to leverage public opinion and action. After being introduced by *The Examiner* as a figure who could 'wider exercise' the values of home, Nightingale had worked with *The Times* to elicit and direct contributions, both essential and luxurious, from the nation's homes to its soldiers. She became the figurehead of the appeal to bring the lessons of the war back home, and, upon her much-anticipated return to Derbyshire, sidestepped the fireworks of the capital and stirred local communities to celebrate the peace in their more familiar confines. Nightingale's role in the war showed that Britain's global activities were taking on a scale and significance that was difficult to explain to a domestic population—yet smaller, homelier motifs, which could be packaged up and understood across regions and localities, could become powerful carriers of national feeling. Nightingale represented home in the middle of a geopolitically complex war whose direct benefit to Britain's national interests was not especially obvious. The incompetent army bureaucracy, by failing to supply the soldiers with comforts, was the perfect contrast to Nightingale's homely work. She was seen to evade uncaring red-tapeism and put the domestic qualities of the household to work for the good of the nation, illustrating that its strength lay in the order and the empathy of its homes and their inhabitants.

Notes

1. Orlando Figes, *Crimea: The Last Crusade* (London: Allen Lane, 2011), xix.
2. L. J. Jennings, ed. *The Croker Papers*, 2nd ed. (London: John Murray, 1885), iii, 354.
3. *The Times*, 29 December 1854, 6; Trevor Royle, *Crimea: The Great Crimean War, 1854–1856* (London: Macmillan, 2004), 502.
4. Andrew Hobbs, *A Fleet Street in Every Town: The Provincial Press in England, 1855–1900* (Cambridge: Open Book, 2018), 126.
5. Stefanie Markovits, *The Crimean War in the British Imagination* (Cambridge: Cambridge University Press, 2009).
6. Holly Furneaux, *Military Men of Feeling: Emotion, Touch, and Masculinity in the Crimean War* (Oxford: Oxford University Press, 2016).
7. Mary Poovey, *Uneven Developments: The Ideological Work of Gender in Mid-Victorian England* (Chicago: University of Chicago Press, 1988), 189.
8. *The Times*, 29 December 1854, 6.
9. Arthur Hugh Clough, *The Poems and Prose Remains of Arthur Hugh Clough: With a Selection from His Letters and a Memoir*, vol. 1 (London: Macmillan, 1869), 217.
10. Figes, *Crimea*, 61.
11. 'The True Story of the Nuns of Minsk', *Household Words*, 13 May 1854, 290–295.
12. 'At Home with the Russians', *Household Words*, 20 January 1855, 535.
13. *The Times*, 5 November 1855.
14. For instance, the reports of the war in *The Times* were read aloud to outdoor crowds in Staffordshire. See Hobbs, *Fleet Street*, 128.
15. Olive Anderson, *A Liberal State at War* (London: Macmillan, 1967), 71, 84.
16. Andrew Lambert and Stephen Badsey, *The War Correspondents: The Crimean War* (Stroud: Sutton, 1994); Markovits, *Crimean War*, 12–25.
17. Markovits, *Crimean War*, 13.

18. Edward B. Orme, 'A History of The Illustrated London News', 1986, https://www.iln.org.uk/iln_years/historyofiln.htm (accessed 22 April 2020).
19. See Yakup Bektas, 'The Crimean War as a Technological Enterprise', *Notes and Records: The Royal Society Journal of the History of Science* 71, no. 3 (2017): 233–262.
20. Natalie M. Houston, 'Reading the Victorian Souvenir: Sonnets and Photographs of the Crimean War', *The Yale Journal of Criticism* 14, no. 2 (2001): 353–383; Helen Groth, 'Technological Mediations and the Public Sphere: Roger Fenton's Crimea Exhibition and the "Charge of the Light Brigade"', *Victorian Literature and Culture* 30, no. 2 (September 2002): 553–570.
21. *The Times*, 13 November 1854.
22. 'Enthusiasm of Paterfamilias', *Punch*, 25 November 1854, 213.
23. Trudi Tate, 'On Not Knowing Why: Memorializing the Light Brigade', in *Literature, Science, Psychoanalysis, 1830–1970: Essays in Honour of Gillian Beer*, ed. Helen Small and Trudi Tate (Oxford: Oxford University Press, 2003), 165.
24. Nightingale to her family, 5 May 1855, in *The Collected Works of Florence Nightingale, Volume 1: Florence Nightingale: An Introduction to Her Life and Family*, ed. Lynn McDonald (Waterloo, ON: Wilfrid Laurier University Press, 2001) [hereafter *CW* 1], 141.
25. George Eliot, *The Mill on the Floss*, ed. Dinah Birch (Oxford: Oxford University Press, 1996), 176.
26. J. Paul Hunter, *Before Novels: The Cultural Contexts of Eighteenth-Century English Fiction* (New York: Norton, 1990), 12.
27. *The Times*, 30 December 1854.
28. John Peck, *War, the Army and Victorian Literature* (Basingstoke: Palgrave Macmillan, 1998), 30; see also Rachel Teukolsky, 'Novels, Newspapers, and Global War: New Realisms in the 1850s', *Novel* 45, no. 1 (1 May 2012): 31–55.
29. Fanny Allen to Parthenope Nightingale, 29 November 1854, Claydon N302.
30. Nightingale to Parthenope, 2 June 1856, in *The Collected Works of Florence Nightingale, Volume 14: The Crimean War*, ed. Lynn McDonald (Waterloo, ON: Wilfrid Laurier University Press, 2001) [hereafter *CW* 14], 410.
31. Arthur Hamilton-Gordon Stanmore, *Sidney Herbert, Lord Herbert of Lea*, vol. 1 (London: John Murray, 1906), 374.
32. *The Times*, 9 October 1854, 8.
33. Quoted in Stanmore, *Sidney Herbert*, 336.
34. Nightingale to Parthenope, 2 June 1856, *CW* 14, 410.

35. Unpublished draft memoir by Parthenope Nightingale, c. 1857, Claydon N389.
36. 'Who Is Mrs Nightingale?', *The Examiner*, 28 October 1854, 682–683.
37. Rachel Bates, '"All Touched My Hand": Queenly Sentiment and Royal Prerogative', *19: Interdisciplinary Studies in the Long Nineteenth Century* 20 (13 May 2015): 25.
38. Lord Raglan to Nightingale, 15 November 1854, *CW* 14, 61.
39. *The Times*, 13 November 1854, 10.
40. 'Nurses of Quality for the Crimea', *Punch*, 11 November 1854, 193–194.
41. *The Morning Chronicle*, 21 October 1854.
42. *The Times*, 24 October 1854, 9.
43. Mark Bostridge, *Florence Nightingale: The Woman and Her Legend* (London: Penguin, 2009), 209.
44. Nightingale to unknown, undated [c. 19 October 1854], *CW* 14, 58.
45. Nightingale to Sidney Herbert, 10 December 1854, *CW* 14, 82, 83.
46. Nightingale to Herbert, 25 November 1854, *CW* 14, 69.
47. Nightingale to Herbert, 8 January 1855, *CW* 14, 109.
48. Nightingale to Herbert, 10 December 1854, *CW* 14, 82.
49. Nightingale to Herbert, 8 January 1855, *CW* 14, 109.
50. Nightingale to Herbert, 28 January 1855, *CW* 14, 126, 128.
51. Nightingale to William Bowman, 14 November 1854, *CW* 14, 64.
52. Sidney Godolphin Osborne, *Scutari and Its Hospitals* (London: Dickinson, 1855), 12.
53. Nightingale to Herbert, 15 December 1854, *CW* 14, 84.
54. Bostridge, *Florence Nightingale*, 225; Nightingale to Herbert, 4 January 1855, *CW* 14, 105.
55. Bracebridges to Nightingale family, 10 November 1854, *CW* 14, 60.
56. Nightingale to Herbert, 25 November 1854, *CW* 14, 69.
57. *The Times*, 24 November 1854, 5; Markovits, *Crimean War*, 49–53.
58. Nightingale to Herbert, 8 January 1855, *CW* 14, 108.
59. Quoted in Edward Cook, *The Life of Florence Nightingale*, vol. 1 (London: Macmillan, 1913), 211.
60. *The Times*, 23 December 1854; 8 February 1855, 8.
61. *The Times*, 12 October 1854, 6.
62. Markovits, *Crimean War*, 25.
63. *The Times*, 13 October 1854.
64. 'The People's Almoner', *The London Review* 6 (21 March 1863): 300.
65. Nightingale to Herbert, 10 December 1854, *CW* 14, 82.
66. *The Times*, 30 November 1854, 8.
67. *The Times*, 25 November 1854, 8.
68. *The Times*, 27 December 1855, 9.
69. *The Times*, 15 August 1856, 6.

70. Nightingale to Herbert, 5 February 1855, *CW* 14, 136.
71. Nightingale to Herbert, 28 January 1855, *CW* 14, 130.
72. Anderson, *Liberal State at War*, 105.
73. J. B. Conacher, *Britain and the Crimea, 1855–56* (Basingstoke: Macmillan, 1987), 21–22; see also Geoffrey Russell Searle, *Entrepreneurial Politics in Mid-Victorian Britain* (Oxford: Oxford University Press, 1993), 89–125.
74. John McNeill and Alexander Tulloch, *Report of the Commission of Inquiry into the Supplies of the British Army in the Crimea* (London: Harrison and Sons, 1855), i; incomplete letter by Nightingale [spring 1856], *CW* 14, 409.
75. Queen Victoria to Lord Raglan, 9 April 1855, Papers of Lord Somerset, London, National Army Museum, 1968–07–280, quoted in Bates, 'Queenly Sentiment', 2.
76. *The Morning Post*, 4 January 1855, 4, emphasis in original.
77. Bracebridge, copied extract from letter, undated, Royal Archives, VIC/MAIN/F/1/80, quoted in Bates, 'Queenly Sentiment', 7.
78. Keeper of the Queen's Purse to Nightingale, 14 December 1854, quoted in Cook, *Life*, vol. 1, 216. For the overlap between Nightingale and Queen Victoria's images see Poovey, *Uneven Developments*, 171.
79. Nightingale to Herbert, 25 December [1854], quoted in *Florence Nightingale: Letters from the Crimea, 1854–1856*, ed. Sue M. Goldie (Manchester: Mandolin, 1997), 59.
80. *The Times*, 25 November 1854.
81. Lady Charlotte Bridgeman, *Journals*, 2 December 1854, http://ladycharlottesdiaries.co.uk/Journal/Select.php?Year=1854&Month=12&Day=2 (accessed 27 April 2020).
82. *The Times*, 8 February 1855, 9; 5 December 1854, 7.
83. Elizabeth Sanderson Haldane, *Mrs. Gaskell and Her Friends* (London: Hodder and Stoughton, 1930).
84. John Vickers to Parthenope Nightingale, 11 December 1854, Claydon N272; English Ladies at Pau to Parthenope Nightingale, 22 December 1854, Claydon N302; Mrs Gallop to Parthenope Nightingale, 2 April 1855, Claydon N319.
85. 'List of Contributions from Derby and Derbyshire to the Hospital Scutari' [1855], Claydon N272; Admiral's Office at Southampton to Parthenope Nightingale, 24 December 1855, Claydon N272.
86. Quoted in Sanderson Haldane, *Gaskell*, 104–105.
87. Quoted in Cook, *Life*, vol. 1, 264.
88. A. W. Kinglake, *The Invasion of the Crimea*, vol. 6 (London: Blackwood, 1885), 393.
89. *The Times*, 24 January 1855, 7.

90. Quoted in Christopher Hibbert, *The Destruction of Lord Raglan* (London: Longmans, 1961), 259.
91. Frederick Robinson, *Diary of the Crimean War* (London: Bentley, 1856), 272–273.
92. Nightingale to her father, 19 August 1855, *CW* 1, 240.
93. Furneaux, *Men of Feeling*, 50.
94. *The Times*, 8 February 1855, 8.
95. *The Illustrated London News*, 24 February 1855. Bostridge, *Florence Nightingale*, 252, mistakenly suggests that MacDonald's account appeared after this engraving.
96. Charles and Selina Bracebridge to the Nightingale family, *CW* 14, 60–61.
97. Cook, *Life*, vol. 1, 238–240.
98. Bostridge, *Florence Nightingale*, 226.
99. Ibid., 253.
100. Mai Smith to her children, 2 September [1855], Claydon N290.
101. Bostridge, *Florence Nightingale*, 253.
102. John Plotz, *Portable Property: Victorian Culture on the Move* (Princeton: Princeton University Press, 2008), xiv.
103. *The Times*, 8 February 1855, 8.
104. See Poovey, *Uneven Developments*, 169.
105. Lytton Strachey, *Eminent Victorians* (London: Chatto and Windus, 1918), 150.
106. Frederick Milnes Edge, *A Woman's Example: And a Nation's Work: A Tribute to Florence Nightingale* (London: William Ridgway, 1864), 6.
107. Quoted in Cook, *Life*, vol. 1, 268.
108. Speech by Lord Stanley, 17 January 1856, quoted in Cook, *Life*, vol. 1, 246.
109. *Derby Mercury*, 5 December 1855, 4, emphasis in original.
110. Ibid.
111. Markovits, *Crimean War*, 189–194; Kate Flint, *The Victorians and the Visual Imagination* (Cambridge: Cambridge University Press, 2000), 76.
112. See Asa Briggs, *Victorian Things* (London: Batsford, 1988), 138.
113. Cook, *Life*, vol. 1, 266.
114. Robinson, *Diary of the Crimean War*, 393–394.
115. Figes, *Crimea*, 478; John Plunkett, 'Celebrity and Community: The Poetics of the *Carte-de-Visite*', *Journal of Victorian Culture* 8, no. 1 (January 2003): 55–79.
116. Bostridge, *Florence Nightingale*, 264, xxi, 265.
117. Rohan McWilliam, 'The Theatricality of the Staffordshire Figurine', *Journal of Victorian Culture* 10, no. 1 (2005): 107–114; Briggs, *Victorian Things*, 124–137.

118. Alfred Bonham-Carter to Parthenope Nightingale, 29 January 1857, Claydon N307.
119. Figes, *Crimea*, 467.
120. Journal of Queen Victoria, 11 March 1856, Royal Archives, Windsor, VIC/MAIN/QVJ, quoted in Figes, *Crimea*, 467.
121. Ulrich Keller, *The Ultimate Spectacle* (Amsterdam: Gordon and Breach, 2001), 216.
122. Ibid., 215–216.
123. *Derby Mercury*, 4 June 1856, 5.
124. Ibid., 8.
125. Julia Smith to William Nightingale, 30 May 1856, Claydon N210.
126. Fanny Wildgoose to Fanny Nightingale, 8 February 1855, Claydon N290; Mary Fox to Parthenope Nightingale, 5 August 1856, Claydon N295.
127. Spencer Timothy Hall, 'The Home of Florence Nightingale', *London Journal* 20, no. 522, 391.
128. Richard Keene to Parthenope Nightingale, 20 August 1856, Claydon N462.
129. Parthenope Nightingale to Richard Monckton Milnes, 12 September [1855], Houghton papers, Trinity College, Cambridge, 18/151, quoted in Bostridge, *Florence Nightingale*, 262.
130. Cook, *Life*, vol. 1, 303.
131. Nightingale to unknown recipient, 12 May 1856, *CW* 14, 402.
132. Nightingale to family, 17 July 1856, *CW* 1, 241.
133. Quoted in Cook, *Life*, vol. 1, 303.
134. Ibid., 304; George Wigglesworth, *Florence Nightingale's Journey Home from the Crimea*, 2010, http://www.wigglesworth.me.uk/local_history/pdf/Florence's%20journey%20home.PDF (accessed 29 April 2020). Thanks to John Slaney for discussions on this point.
135. Parthenope Nightingale to Lady Canning, 2 September 1856, Claydon N23.
136. *The Morning Chronicle*, 13 August 1856.
137. *The Times*, 15 August 1856.
138. Bostridge, *Florence Nightingale*, 305–306.
139. Nightingale to the female tenantry of Pleasley, 14 August 1856, *CW* 14, 441.
140. Nightingale to Parthenope, 9 July 1855, *CW* 14, 198; William Nightingale to Fanny Nightingale [1855], Claydon N306.
141. William Nightingale, private note, undated, Claydon N73.
142. William Nightingale to Fanny Nightingale, 1 May 1855, Claydon N306.
143. Hilary Bonham-Carter to Parthenope Nightingale [1856], Claydon N56.
144. Selina Bracebridge to Parthenope Nightingale [1856], Claydon N54.

145. *Derby Mercury*, 3 September 1856, 3.
146. Nightingale, private note, August 1856, in *The Collected Works of Florence Nightingale, Volume 2: Florence Nightingale's Spiritual Journey*, ed. Lynn McDonald (Waterloo, ON: Wilfrid Laurier University Press, 2001) [hereafter *CW* 2], 390.
147. Nightingale, private note [1856–7], *CW* 2, 391–395, emphasis in original.
148. Nightingale, *Notes on the Health of the British Army*, *CW* 14, 587.

CHAPTER 7

Working from Home

Florence Nightingale returned from the Crimea in 1856 as a national heroine. Yet, rather than triumphant, she was shaken and depressed by the death and suffering she had seen in the war hospitals and profoundly aware of the limited impact of her work. To Nightingale, the lessons learned from Scutari were less a cause for celebration than an urgent signal for broader, more systemic reforms of army sanitation. Nightingale's mood was sombre, but perhaps even more definitional to her character at this time, she was also physically depleted. For most of the rest of her life, Nightingale experienced muscular weakness, exhaustion, depression of mood, and spinal pain, symptoms that confounded medical experts and members of her family. It was only late in the twentieth century that brucellosis (probably caught through drinking infected milk while in the Crimea) was retrospectively diagnosed as the likely cause of her chronic condition.[1] This illness ebbed and flowed in intensity for the next forty years of Nightingale's life. A first, relatively brief, collapse in the Crimea in 1855 was followed by a more severe physical breakdown in 1857, which caused Nightingale to be invalided almost permanently until 1881. A particularly debilitating attack during Christmas 1861 marked her lowest point health-wise. Between 1861 and 1867, she was unable to walk even short distances and was entirely bedridden.[2] It was only after 1881, when Nightingale was over sixty, that she started to appear in public again, but even then, only selectively and briefly.

P. Crawford et al., *Florence Nightingale at Home*,
https://doi.org/10.1007/978-3-030-46534-6_7

Forced to adjust to these new restrictions on her activity, Nightingale absented herself from public view, but not from public life, instead developing new routines and working methods. Subverting norms associated with invalidism, Nightingale, within her home, worked prodigiously on British nursing reform and public policy (Chapters 4 and 5). What has been less well documented is that she also turned her attention overseas to the British Empire, becoming intensely involved with sanitary reform in India.[3] Although India formed the most substantive body of her engagement from London, Nightingale also worked on other imperial contexts, albeit not with the vigour and persistence she devoted to the subcontinent. She wrote papers on the aboriginal races in Australia, for example, and was an active campaigner to ensure their survival.[4] She also oversaw, albeit with variant success, the deployment of nurse training schemes to other outposts of the British Empire, most prominently to Australia (1868) and Canada (1875).[5]

The interest in India is not hard to explain. After the Crimea, India presented the next site of intense political concern overseas. Within a year of Nightingale's return, Britain's supreme colonial possession became home to the most significant imperial emergency the British government had yet faced. In an uprising which has become known variously as the Indian Mutiny or the Sepoy Rebellion of 1857, Indian soldiers rose up *en masse* against their British commanders, joined by local rulers and thousands of ordinary Indians in a bloody struggle that threatened to destroy British colonial power.[6] So soon after witnessing the appalling conditions with which British troops in the Crimea had had to contend, similar fears surfaced regarding the conditions of the British military in India. In short, addressing the emergency in India became the next priority of national importance for the British government and created a logical entry point for Nightingale to channel her interests.

It is this curious relationship between Nightingale's new secluded homebound lifestyle and her increasingly expansive outward-looking concerns in India that forms the focus of this chapter. Nightingale may have been confined within her home due to illness, but she used this time to work towards improving, first the conditions of soldiers, but eventually also the homes, health, and prospects of ordinary Indians. She did this in ways that were both typical and atypical of the time. Empire, although distant and representative of unfamiliar climates, geographies, peoples, and customs, was nevertheless regarded as an extension of the

British homeland. Public health, furthermore, was seen as a key tool to disseminate Western civilisation and progress.[7] In proffering advice for Indian imperial citizens, Nightingale was integrally participating from her London home in her national duty. Her approach tacitly accepted many of the central assumptions of colonial paternalism, yet was also enlightened and exceptional for the time.

Brucellosis and New Ways of Working

In one of the many paradoxes that came to characterise her life, brucellosis sequestered Nightingale within various houses, usually propped up in bed or on a chaise longue within the confines of a single room. Isolated by her condition, she found herself having to spend much of her time in the kind of domestic cage that she had railed against as a younger woman. This post-Crimean imprisonment within the home differed in several crucial ways, however, from the restrictions she had described her family homes as representing before 1853 (Chapter 3). Now, it was the walls that physically contained her, rather than the social mores and expectations associated with being a middle- or upper-class lady in polite society. Poor health may have dictated that Nightingale was not able to leave her house easily, but since the Crimea her reputation was secured, and she was able to determine her own daily routines and preferences—and most importantly working life—in ways that she had not been able to command as freely as a young woman. Nightingale nevertheless returned to earlier metaphors of the home as a prison as she reflected on her state, complaining in 1861: 'I have passed the last four years between four walls, only varied to other four walls once a year; and I believe there is no prospect but of my health becoming ever worse and worse till the hour of my release', and similarly repeating in 1866 that her illness had rendered her 'entirely a prisoner to my bed' (Fig. 7.1).[8] This new sedentary reality of Nightingale's life saw her also physically change in ways which were increasingly far removed from the young attractive images that circulated in the public domain and underpinned her celebrity status. So rarely did Nightingale appear in public post-Crimea, and so forceful was the public appetite to hear of the gentle ministrations of the lady with the lamp tending to the sick and wounded soldiers, that the reality of her new housebound existence almost did not matter. Indeed, by the end of the century, the Nightingale myth dwarfed her reality to such an extent that people were said to be 'shocked to learn ... [that she] was still alive'.[9]

Fig. 7.1 Photograph of Florence Nightingale in bed, in her room at 10 (formerly 35) South Street, London. Elizabeth Bosanquet, 1906, reproduced in Edward Cook, *Life of Florence Nightingale* (London: Macmillan, 1913), vol. 2, 307

To those close to her, however, Nightingale—true to form—was a vocal patient. She often described to others the painful and incapacitating nature of her state and was not shy in predicting her own demise. As early as 1857 (aged just thirty-seven), debilitated through exhaustion and chronic weakness, the severity of her second attack of brucellosis—she thought it might be typhus—caused her to draw up her will, writing to her friend and political ally Sidney Herbert a letter that she placed in an envelope marked '[t]o be sent when I am dead'.[10] This presumption of her imminent mortality was shared by members of her family, with Parthenope also confessing her fear that 'she [Florence] will not live long' when she wrote to Elizabeth Herbert in the weeks following her sister's return from the Crimea.[11]

Despite these concerns, Nightingale powered on. Indeed, contrary to all expectations, the period between 1856 and the mid-1890s was to become the most industrious of her life, despite (or perhaps because of) her seclusion and confinement.[12] The way this new era was characterised by 'almost uninterrupted activity' was not lost on contemporary commentators: Millicent Fawcett commented that 'few people, even of the most robust frame, have done better and more invaluable work'.[13] Or as pithily put by Lytton Strachey: 'her real life began at the very moment when, in the popular imagination, it had ended'.[14] In some ways, this was not surprising. Nightingale hated idleness of any kind and her Crimean experiences had shown her the gargantuan size of the work that needed to be done if she were to effect widespread change.[15] Contemporary social norms dictated that chronic illness suffered by a middle- or upper-class woman would normally involve a life sentence of rest and passivity, but for Nightingale her illness became a freeing force, rather than a confining one.[16] Nightingale made her sickroom her office and creatively deployed invalidism as a means to wield considerable influence over public policy, while remaining, on the surface at least, politically unaffiliated and non-threatening.[17] Illness allowed Nightingale to restrict and curate her audiences (including her mother and sister), keep herself from public gaze, and undertake her work undistracted.[18] As Miriam Bailin notes, 'invalidism permitted the inversion of this sentence imposed by her gender and class by permitting her to sequester and immobilize herself while labouring prodigiously on projects of both national and imperial importance'.[19] Nightingale can be seen as having actively orchestrated this seclusion. She had already written in 'Cassandra' in the early 1850s that in a world dominated by social requirements 'bodily incapacity is the only apology valid'.[20] Following her own logic, after returning from the Crimea Nightingale regularly cited her illness as the reason for turning away impromptu visitors, postponing meetings or putting people off entirely. Likewise, she invoked it as the reason why she had to organise her time quite rigidly—with scheduled visitors given a half-hour slot via written invitation, so that she might not become too exhausted.[21] Nightingale also used her sickness to add an urgency to her appeals, frequently citing her ill health as a reason why actions should be undertaken quickly.[22] Her biographer Mark Bostridge was careful in his wording when he described how Nightingale imposed a 'rule of strict seclusion upon *herself*'.[23] Even in later life, once her symptoms had somewhat subsided, it was clear how integral being ill had become to Nightingale's identity and that 'invalidism formed part of the grain of her life'.[24]

It was during this housebound second-act of her life that Nightingale turned her attention to India, while simultaneously continuing her work developing British nursing and on sanitary reform more generally. This new interest gradually broadened from a quite narrow focus on the hygiene and sanitation of the British military in India into the wider domain of social reform, eventually encompassing campaigns on education for women, agricultural reforms, famine relief, and Indian independence. Her new mode of working on these Indian topics, as well as on her continually expanding home interests (Chapter 5), differed from how she had worked before the Crimea. Whereas previously Nightingale had been at the coalface witnessing first-hand the issues she was keen to address, her post-1856 reform work was all conducted from her bedroom. India may have become a new fascination, but Nightingale never actually stepped foot in the subcontinent. It took some time for her to come to terms with the obvious restrictions necessitated by her immobility. Nightingale's regret about her remoteness, particularly in the early years of her India work, was readily apparent. As early as 1857, when the news arrived in the UK that the Indian Mutiny had broken out, she wrote to Lady Canning, the wife of the Viceroy, to express her desire to 'come out, at twenty-four hours' notice, to serve in any capacity'.[25] This offer—involving a forty-day sea passage—was entirely unrealistic, given she had just had a severe attack of brucellosis the month before. Nonetheless, Nightingale on several occasions reiterated her desire to work in India.[26] In 1865, she wrote, 'if there were even any hope of my reaching India alive, and of my being able to go on working when there, as I do here, I believe I should be tempted to go'.[27] Even as late as 1888, when Nightingale was approaching seventy, discussions over India led her to lament that she wished she 'were going too'.[28]

The practically minded Nightingale, used to influencing change, developed working methods suitable for her unique brand of bedroom imperialism. Neither the geographic nor the contextual remoteness of India were going to silence her impulses to advise and reform. As Gérard Vallée has described, Nightingale's method of working was to read everything, collect evidence, discuss with experts, and lobby people of influence. She did this through dedicated study, receiving visitors to discuss pertinent issues and, perhaps most importantly of all, correspondence.[29] Letters gave Nightingale an entry point from the private sphere of home into public participation, allowing her to fulfil her political aspirations via a medium which was deemed socially acceptable. As Chieko Ichikawa argues, Nightingale's letters enabled the 'coalescing of her masculine ambition into feminine duties'.[30] From the busy base of her London home, regularly visited by the great and good, Nightingale was the queen bee around whom her adoring public, and others with social reforming agendas, hummed. Her abundant correspondences were the means through which her ideas for change became pollinated, often travelling well beyond her immediate circle.[31]

Nightingale's London Homes

If, as Judith Flanders has argued, 'home created family life', Nightingale, choosing to live alone without a family, devised arrangements within her own home that were unusual in numerous ways.[32] As was always the paradox with Nightingale, however, in several crucial elements, she nevertheless conformed with contemporary social practices. Despite her attention to setting up institutions, Nightingale's own sickness was played out at home, as was entirely typical for members of middle- and upper-class British society. She famously visited Queen Victoria and Prince Albert at Balmoral in autumn 1856 to tell them personally about the reforms to army sanitation that she wanted to see tackled in a Royal Commission, but thereafter her ill health impeded her to such an extent that she rarely left London.[33] After her second major collapse, Nightingale travelled to Malvern to take the hydrotherapy there in 1857 and 1858 (making a return visit in 1868). From 1866, she also intermittently visited Embley Park (until her father's death in 1874), Lea Hurst (between 1867 and 1880, especially while nursing her elderly mother in the 1870s), and her sister Parthenope's home, Claydon House. Other than these odd forays away from the capital, Nightingale remained largely housebound between 1857 and 1881, after which time her physical health improved slightly, and she started to make occasional public appearances again.

Nightingale took some time to settle in London, but it was clear to her that to undertake her work properly she would need to situate herself close to the politicians through whom she worked. Before arriving in her private house at 35 (later 10) South Street in 1865, which was to remain her home until her death in 1910, Nightingale stayed at a number of places, often summering in rented accommodation in Hampstead or Highgate.[34] Until her final South Street move, her principal London residence was the Burlington Hotel in Westminster, where she lived first in the main body of the hotel and then latterly—between 1858 and 1861—in the hotel's annexe, which contained several private apartments for long-term residents.[35] The flurry of activity that emanated from these Burlington headquarters was such that they soon became warmly known as the 'little War Office' or the 'little India Office'. From here, Nightingale worked closely with Sidney Herbert, 'her public voice', the Secretary of State for War (1859–1861), and a dear friend. They met almost daily until his death in 1861, with the tightness of their association tellingly revealed in their affectionate nicknaming of their Burlington group as 'our cabal'.[36] Initially, Nightingale's mother, Fanny, and sister, Parthenope, accompanied her to the Burlington, ostensibly in the role of offering support. In a typically abrasive dismissal of their intentions, Nightingale reflected on their actions, finding them confining and distracting and, ultimately, more of an irritant than an aid. 'What have my Mother & Sister ever done for me?' she recounted. 'They like my glory – they like my pretty things. Is there anything else they like in me?' It was not lost on Nightingale that their lack of support for so many years contrasted starkly with their desire to support her now she had achieved widespread fame.[37]

After Herbert's death, Nightingale could no longer bear the Burlington where she had worked with him so closely and moved away for a change of scene: first to a rented house in Hampstead and then to the London house of her sister's husband (Sir Harry Verney MP) at 32 South Street in Mayfair. With their drastically different lifestyles—Parthenope's increasingly centred on the circulations of polite society since her marriage, and Nightingale's focused around work—it was never going to suit Nightingale to co-habit, so when a house became available just a few doors down the street, she jumped at the chance to take her own accommodation. Despite her aspirations towards independence from her family, these moves within London, which involved transporting her considerable books and papers, redecorating and making residences comfortable, were entirely dependent upon her father's financial largesse.[38] In fact, it is important to note that despite her constant work, Nightingale was not paid for anything she did and was financially dependent upon her father, who from this period covered all her living expenses in addition to granting her a £500 allowance each year.[39]

Playing into the conventional mores of the time, and directly reflective of her own formative experiences at Lea Hurst and Embley Park (Chapter 2), Nightingale was very concerned with the efficient running of her home. As was typical for a woman of her status, Nightingale maintained a household of four or five servants, including a personal maid who for many years was a woman called Temperance Hatcher, whom Nightingale clearly held in high regard as she was remembered in Nightingale's will.[40] Evidence seems to suggest that Nightingale was a demanding mistress, although also generous in her renumeration of her employees.[41] She routinely employed a cook, a maid, and a messenger, but rather than leaving them to their own devices, she could not help but micro-manage their daily business, performing precisely the gender-bound household management role from which she had struggled so hard to be free. A stickler for detail, Nightingale reviewed the cooks' menus, purchases of bed linen and food, and personally organised the detailed entertainment schedules of her visitors. She frequently asked her relatives and neighbours to perform social duties on her behalf and leaned on her family to entertain her guests at Embley or Lea Hurst.[42] While shunning the social conventions of family, a picture nevertheless emerges of Nightingale also deeply reliant upon them. As part of her way of working during this period, she retained a close confidante at all times. Acting as the Victorian equivalent of her PA was first Parthenope who, among other things, managed public donations during the Crimean War (Chapter 6), and then her aunt Mai Smith (between 1857 and 1860) who worked simultaneously for poet Arthur Hugh Clough (married to Nightingale's cousin Blanche) and acted in a secretarial role (1857–1861). Later, her cousin Hilary Bonham-Carter (between 1860 and 1862) saw to Nightingale's affairs, until the role was more durably taken over by Dr John Sutherland (from 1861 through to the 1880s), who as well as being one of the few remaining members of her early Burlington 'little War Office' was also Nightingale's personal physician.[43] The tedious necessity of managing domestic duties was only an acceptable part of her life, however, when conducted on her own terms. While Nightingale was happy to use family connections, or money, to help oil the wheels of her reforming ambitions, she also complained of the burden of family when duties concerning them passed to her. Nightingale had spent some eight months at Embley looking after her parents in 1872, such that 'my London work was almost ruined'.[44] While seeking carers for them in 1873, she complained that 'upon me, the only one of the family who has any real work to do, the whole thing is thrown'.[45]

Victorian households were typically divided into areas that were private (bedrooms, study) and public, where visitors would be received (drawing rooms, dining room, morning room), but Nightingale blurred these conventional boundaries by making her bedroom her office and her home her headquarters.[46] This preference did not come out of the blue, but was the extension of the working preferences that she had developed at Scutari, where she would work at a desk in her bedroom—her aunt (Mai Smith) remarking of this arrangement that she had never seen 'a mind so continuously concentrated on her work'.[47] This decision to work at home not only subverted the Victorian trope of the passive female invalid, it also allowed Nightingale to make the point that her life was devoted to her work, not to social conventions. When she redecorated her apartment in the Burlington Hotel annexe in 1858, her father sent her a few things to help furnish her drawing room. Nightingale's response was one of indignation, curtly reminding him that she did not have a drawing room and explaining that such a place was symbolic of 'the destruction of so many women's lives'.[48] These views must have softened, however, as her final resting place, 10 (formerly 35) South Street, did contain precisely this sort of formal reception room. Although she was never one to care too much about the accumulation of material objects, as evidenced by her preferences expressed in her final will, Nightingale did care about the good quality of her furnishings.[49] Following her own model of how domestic spaces should be organised for optimal health—with good ventilation, minimal dust, and just the bare essentials—her London suite seems to have been more sparsely furnished than the more luxurious surroundings that she knew from her earlier life.[50]

In line with this preference for simplicity, Nightingale liked plain, well-cooked food. Bostridge lists her favourite recipes as including boiled mutton and turnips.[51] Once in London, she also regularly asked her mother and sister to send her 'boxes' of local produce from Hampshire, evidently finding this to be of a quality superior to what was available in the capital. She estimated in 1870 that these edible home comforts—many of which she gave as gifts to nurses and other colleagues—cost her approximately £150 a year.[52] She also was said to enjoy ginger and port wine in moderation.[53] After a day's hard work in bed, or on her *chaise longue*, Nightingale customarily ate alone. Her only indulgence was her beloved cats. It was rumoured that at one point she had as many as seventeen, a number of whom were referred to by name in her letters with their paw prints christening many of her personal papers.[54]

Nightingale's London life was both typical and exceptional for her time. On the one hand, she acted with the confidence of an astute politician, keen to situate herself at the heart of things, advocate for reform, and fearless in whom she approached. Unmarried, critical of family structures, social rituals, and expectations, her unique brand of home life successfully challenged stereotypes of both femininity and invalidism. Rather than retreating to a life of inaction, corporeal weakness instead created for Nightingale an opportunity to finally excuse herself from the niceties of polite society and get on with more work than ever. Nevertheless, the central fact remained that her physical location was locked within the home, rather than in the workplace or among the people she hoped to help, and thus centrally complied with dominant social norms. For all her exceptionalism, Nightingale was ultimately a dependent, deeply embedded in her family networks and reliant on the financial support of her father. Furthermore, her home management (from decoration to assiduous oversight of the staff) very much fitted the mould of the typical middle-class mistress. This central tension, which saw Nightingale constantly navigating a fine line between novelty and tradition, characterised her entire life but was brought to particular prominence when her health precluded her from participating in any more physical nursing. Unsurprisingly, this unusual combination of radicalism and social conservatism also characterised the work she undertook on India.

Nightingale's Work on India

In Victorian Britain, empire, with its unfamiliar populations, hot climates, and dramatic tropical diseases, denoted difference—a difference, furthermore, that became increasingly interpreted as inferiority as the nineteenth century progressed. Indigenous societies were routinely compared to the so-called civilised benchmark provided by the industrialised nations of the European colonisers. When Joseph Chamberlain, head of the Colonial Office, stated in 1895 that he believed 'the British race is the greatest of the governing races that the world has ever seen', he was articulating his supreme confidence in his countrymen to guide, educate, and lead by example as part of his philosophy of 'constructive imperialism'.[55] India was regarded as the jewel in the British imperial crown, not only because the British had held a presence in India since 1600, but because it evocatively represented—at least on the surface—all that was 'right' and worthy of celebration in empire. The British public were fed exotic imagery of maharajahs, tigers, and elephants, while the East India Company's 1850 gift to Queen Victoria of the jaw-dropping Koh-I-Noor diamond was given extensive press coverage, and new Indian goods flooded British markets and became markers of luxury and worldliness in sophisticated British homes.[56] Despite her predilection towards avoiding ostentation, Nightingale, working alone in her bedroom on such prosaic matters as sanitation, was undoubtedly also influenced by this popular enthusiasm and assumption of imperial purpose. Her work took her in directions that exposed the gulf between these lavish popular portrayals and the poverty-stricken reality of most Indian lives, but in so doing also played directly into a domain in which it was acceptable for British women of her generation to publicly oversee philanthropic acts.[57] From the 1860s onwards, many middle- and upper-class Englishwomen became interested in India as 'a glorious field' in which they 'could prove their contributions to the imperial civilising mission'.[58] British women philanthropists, as guardians of the home in the UK, became cast as key in this territorialising project: able to provide the peoples of empire with the necessary expertise and insight as civilisers, domesticators, and educators.[59] It was widely perceived that '[t]he successful running of the empire required womanly skills of household management'.[60] These common socio-cultural presumptions that saw women as central in empire-building initiatives also would have framed Nightingale's thinking in the 1860s and 1870s, when she became very supportive of female emigration to Australia and Canada. The key to establishing an effective sanitary environment in such far-flung colonies lay in ensuring that nurses were an integral part of these new societies.[61]

Nightingale's agenda for India, initially at least, was shaped by her experiences of nursing during the Crimean War. Her shock at the inadequate provision for the health of the British army in the Crimea drove her work with Sidney Herbert as soon as she got back to England, setting up a Royal Commission to examine what could be done better for military health in future. The *Royal Commission Appointed to Inquire into Regulations Affecting the Sanitary Condition of the Army, the Organization of Military Hospitals, and the Treatment of the Sick and Wounded* (1857–1858), often referred to simply as Nightingale's First Royal Commission, became the direct model for her Second Royal Commission (1859–1863) that extended her vision to look at similar themes specifically within the Indian military and (eventually, as she lobbied for the brief to be extended) also the surrounding communities.

Information for the Indian Commission was laboriously gathered and consisted of responses to questionnaires composed by Nightingale that were sent out to various stations in India and then combined with other local evidence from officials on the spot. It is testament to the Herculean task that Nightingale undertook collating such vast amounts of information that the final report, mostly authored by Nightingale, was over 2000 pages long.[62] The report was not received with the fanfare she had hoped for. Its heavily critical tone undermined efforts to construct a positive image of the British Empire, and perhaps most importantly, the far-reaching sanitary reforms that it recommended appeared a considerable potential drain on imperial finances. Regarded as too long to be easily digestible, the report was never widely circulated. Instead, Nightingale was asked to comment on the Commission's findings, which she did in a shorter publication that became known as *Observations* and was distributed throughout the cabinet and the India Council.[63] Although eventually the recommendations of this vast survey did become the basis of the colonial sanitary improvement plans for military barracks, it was one of Nightingale's most frequently vocalised regrets that change took so long to materialise in India, despite her painstaking presentation of the evidence. In 1888, she lamented how the country had 'neglected for twenty-five years to adopt measures recommended by the royal commission'.[64] Offering a summary of the Indian Commission's findings in a paper presented to the National Association for the Promotion of Science held in Edinburgh in 1863, 'How People May Live and Not Die in India', Nightingale blamed the disease problems of British soldiers squarely on remediable issues of sanitation, challenging the dominant thinking of the time that tended to blame ill health experienced in colonial locations on unfamiliar tropical climates.[65] Ever practical, Nightingale realised that adequate infrastructure and personnel were needed to ensure the implementation of her sanitary plans. To this end, she promptly participated in the appointments of a cohort of sanitary commissioners to serve in the Presidencies of India, who were tasked with collecting mortality and morbidity statistics in both the military and civilian populations. She also spent considerable energy devising a detailed plan to reform nursing in Indian hospitals, which involved sending graduates from the Nightingale School to India to start similar nursing schools.[66] After several years of exerting epistolary pressure on the Government of India to take up such a plan, the scheme's proposals were ultimately extended and reformed in ways not acceptable to Nightingale, and finally rejected by Sir John Lawrence (Viceroy of India, 1864–1869) in 1867.[67] Nightingale was to return to this issue later in her life, but never saw her nurse training scheme for India come to fruition.[68]

During the late 1860s, however, Nightingale's preoccupations appreciably widened and, most notably, she started to work with Sir Bartle Frere, former Governor of Bombay and President of the India Office Sanitary Committee, to look at the health condition of India more generally. Famines in India between 1861–1867 and 1876–1879, as well as the devastating cholera epidemic that swept the subcontinent between 1867 and 1877, showed Nightingale just how widespread and urgent threats to human life in India were. Correspondingly, her focus shifted from health conditions within army barracks to the communities in their immediate vicinity, and she started to think in detail about structural health reforms in India more widely. Nightingale became an honorary member of the Bengal Social Science Association in 1869 and a key agitator behind the Bombay Village Sanitation Bill. It took several years, but by 1888 one of her most tangible successes was that sanitary boards were established in every Indian province.[69]

Recognising that sanitary improvement relied on infrastructural development, Nightingale time and again returned to the central importance of adequate sewerage and drainage both in the villages and in the surrounding fields. Consequently, irrigation became a central preoccupation. Her 1874 publication 'Life or Death in India' ended with an appendix entitled 'On Life or Death by Irrigation', which began with the powerful claim that '[w]herever water for irrigation and navigation exists, famine is effectually met'.[70] Once she had started working in the new realms of agricultural reform, it was as if the floodgates had opened. For the rest of her career, Nightingale's horizons became much wider and her ideological intentions more complex, as she suggested structural improvement in India, particularly in the realms of land tenancy, agricultural practices, and education. It was also during these final decades of her life that she became increasingly interested in the burgeoning movement for Indian independence. By the 1880s, she was regularly receiving visits from Indian delegates, including representatives of the Indian National Congress (INC), the Indian Association, the All Indian Muslim League, the Bombay Legislative Council, and the East India Association. It is not well known that by 1883 Nightingale had openly declared her support for the INC in a published paper, 'Our Indian Stewardship', and for the rest of her life was an ardent supporter of self-rule, regularly granting audience to prominent activists such as Narayan Ganesh Chandavarkar, Lalmohan Ghose, Gopal Krishna Gokhale, Behramji Malabari, Salem Mudaliar, Dadabhai Naoroji, Mary Scharlieb, Prasanna Sen, and Dinshaw Wacha.[71] Evidence exists showing that Mahatma Gandhi himself held Nightingale in high regard, although he praised her dedicated character and her British reforming work, rather than specifically her work in the Indian context.[72]

Empire as Home

Home was a recurrent metaphor running through the discourses of empire with which Nightingale worked. Its rhetorical deployment was flexible, shifting to suit different situations, but two leading usages can be discerned. First, the word home powerfully tied imperialists to the mother country, with all the connotations of comfort, familiarity, and safety that Britain represented. When imperialists contrasted their colonial outposts with home—often describing their lives in the colonial locale as being lived in exile—they were underscoring the insurmountable difference between the civilised modernity from which they came and the unhomely, unfamiliar spaces of empire into which their careers had thrust them.[73] Contrastingly, however, home was also used as a way of expressing ownership of the empire, by stressing unity. As succinctly articulated by Nightingale, 'England is <u>one</u> wherever her people dwell'.[74]

Distant outposts were selectively referred to as home and domesticating language was used, particularly in political contexts as a means of familiarising geographically remote places and emphasising shared cultural understandings and allegiances between colony and homeland.[75] As Tony Ballantyne and Antoinette Burton have shown, for all the difference it espoused, the British Empire of the nineteenth century was ultimately a 'territorializing' project, a 'place-making' regime 'whose cultural forms ... add up to a historically particular global model of "culture" and "civilisation"' that the West attempted, with variant success, to export to these territories.[76] India may have seemed alien in many ways, but the modern technologies of railways, steamships, and the telegraph from the 1860s onwards helped to rescale time and place and make it more relatable as a distant part of the home nation.[77] It was home to unfamiliar people, geographies, and cultures, but even the most reluctant imperialists embraced it as a serious responsibility, an annexe to the home country that could be managed through careful reforming work.

Nightingale was born not only into the world of empire, but also into a time and class context where ideas of bourgeois respectability were particularly powerful. The combination created a potent merging of domestic ideas of respectability with ways to govern.[78] Etymologically, it is no accident that the verb to domesticate (*dominus*) is connected to domination (to be Lord of the *domum*). Domesticating empire was not only part of the political project of cementing European claims to ownership, it was also part of the ideological claim that non-Western cultures were inferior, and assumed that the Western domestic progress narrative was universal and incontestable.[79] The nineteenth-century Western gospel of hygiene became synonymous with modernity, linking, to use Ruth Rogaski's turn of phrase, issues as prosaic as 'where people went to the bathroom', with ideologies as momentous as how citizens 'envisioned the nation'.[80] Public health reform, rather than being a benign gift to colonised peoples, was a spatial form of European governance—one that was selectively deployed, disruptive to indigenous cultural practices and community arrangements, and was often predicated upon a central assumption of the superiority of the West. Just as Mary Poovey has argued in relation to Nightingale's health advice to British working classes (Chapter 4), reforming advice in India might also be critiqued as a means of class surveillance and regulation, albeit 'by means of known laws of cleanliness rather than by military force'.[81] An 1864 Nightingale letter to the then newly appointed Viceroy

Sir John Lawrence is revealing: he needed to conquer 'India anew by civilization', she urged, by 'taking possession of the empire for the first time by knowledge instead of by the sword'.[82]

Nightingale saw homes as central to promoting hygienic modernity in India.[83] Unsurprisingly, this advice directly reflected the preoccupation that she outlined in *Notes on Nursing* (1860), which recommended changes within the private spaces of each individual home as the key means of fundamentally changing attitudes at a grassroots level (Chapter 4). Generally speaking, health advice and schemes that had worked in Britain were assumed to be exportable from the UK context to the Indian one.[84] Light and the free circulation of fresh air were particularly important symbols of civilisation for Nightingale that formed the backbone of her recommendations for barrack and hospital design in India, despite the drastically different climate, prevalent diseases, and entomology that the country presented.[85] Similarly, colonial suggestions mirrored health models devised at home, for example Nightingale's proposals in the second Royal Commission report for the implementation of a health governance system (based on British health boards and sanitary commissioners); her outline of a nurse training scheme for India; her advice on improving Indian village sanitation; and her scheme of health missioners for rural India. The traffic of expertise did not only flow one way, however, and sometimes issues pursued in the colonial context prompted Nightingale to think about equivalent sensitivities in the UK. In 'Rural Hygiene' (1894), for example, she considered how attempts to communicate health schemes to an Indian population—one of which depended on trained lecturers working in parallel with 'instructed native women visiting and teaching health habits to the village poor native women in their own homes'—might help address similar challenges in Buckinghamshire villages (Chapter 4). 'Let not England lag behind', Nightingale warned, in an intriguing reversal of the conventional imperial hierarchy.[86]

Nightingale regularly used the language of home in her framing of Indian-based reform, stating that she had worked on the subcontinent 'so much' that time spent on this always felt as though she was 'going home'.[87] Helping Indians, furthermore, was not an act of distant philanthropy to culturally remote foreign people. It should rather be understood, said Nightingale, as the British helping their 'fellow subjects'.[88]

Nightingale the Empire Builder?

While deeply embedded in a nation that was enthusiastic about empire, Nightingale nevertheless held unusually progressive views. As Vallée has shown, Nightingale's tireless Indian work was conducted 'against a strong stream of British opinion' and she was well ahead of her time, especially in the way she campaigned for sanitary, educational, and land reform for ordinary Indian citizens in situations yielding no obvious immediate benefit for European colonisers.[89] Throughout her life, Nightingale was a devoted Liberal and often concerned for those who had little voice in society.[90] She saw herself as the champion of the Indian peasant (*ryot*) and mounted an energetic campaign against the revenue collectors (*zemindars*) who challenged the peasants' rights to occupancy of their land.[91] She reacted with great sensitivity to the dire repercussions of the Indian Great Famine (1876–1878), describing in heartfelt detail the 'little naked living skeletons' of child victims and vehemently criticising the 'slavery' in which many *ryots* were positioned in relation to their rapacious *zemindar* overlords.[92] Nightingale's solutions went to the heart of radical structural reform, suggesting, among other things, that *ryots* should be enabled to pay off their debts and should have loans made available to them on special terms. Systems of land assessment, furthermore, should be structured in ways that gave the potential for the *ryot* to ultimately be independent, giving them the opportunity to rise out of the cycle of poverty and subjugation in which they found themselves.[93]

Also unusually for the time, Nightingale was explicit in her sensitivity to Indian cultural traditions. As early as 1853, she complained that Lord Cornwallis, the eighteenth-century Commander in Chief of British India, had acted with supreme cultural arrogance as if 'the Hindus had no old feelings, traditions and customs'.[94] This central belief became further embedded during the post-1857 years of her active Indian involvement, and it was one of the hallmarks of Nightingale's later health and social reform advice for India that indigenous cultures should always be respected. In the 'Health Lectures for Indian Villages' (1893), for example, she sought to reassure the editors of local vernacular journals over their concerns that sanitary work would be undertaken only 'in accordance with European ideas'. Nightingale was at pains to stress that her scheme was not intended to disrupt national customs. 'On the contrary', she insisted, the aim was to work 'through the villages themselves' with respect paid constantly to local religion and practices.[95] In

Nightingale's model, health lectures and demonstrations were not to be delivered by Europeans, but rather by members of local Indian Associations who, she pointed out, better knew 'the habits of the people and [were] able to sympathize with them and help them, without offending their prejudices'.[96] This was an educational model of self-help very similar to that which she had presented in the British context, one that foregrounded local cultural relevancy while also acknowledging the need for some 'great central authority' to enforce and regulate provision.[97] Also echoing her vision for what district nurses and health visitors might achieve in the British context (Chapter 4), Nightingale saw women in India as being key to on-the-ground social improvement. Her rural health education schemes, rolled out from the mid-1880s, all presumed the central importance of Indian women as the facilitators of this knowledge dissemination. Ever sensitive to local dynamics, Nightingale advised: '[i]t is a truism to say that the women who teach in India must know the languages, religions, superstitions and customs of the women to be taught in India'.[98]

Notwithstanding this cultural awareness, however, Nightingale never quite separated her agenda in India from assumptions about the innate applicability of European moral guardianship.[99] To use Mary Poovey's powerful characterisation: '[t]he crusade that Nightingale originated … was explicitly colonial. With the complacency of an imperialist, Nightingale assumed that bourgeois domesticity and cleanliness were universally desired'.[100] Most historians concede this tendency when describing Nightingale's attitudes during the 1850s and early 1860s. After this point, however, more sympathetic historians and biographers tend to portray Nightingale as having dramatically evolved, stressing her subsequent move away from overt imperialism, towards a position that prioritised helping Indians on their own terms, with a view to self-governance.[101]

The early empire-building rhetoric is relatively easy to locate within Nightingale's first Indian writings, particularly when she was responding to the crisis of 1857. Her points of reference can be seen as directly stemming from language deployed in the first Royal Commission. Nightingale evocatively recalled the horrific scenes she had witnessed at Scutari, with men that she compared to 'living skeletons, devoured with vermin, ulcerated, hopeless and helpless'. If this was not avoided in future, she asked if 'the next thing we shall hear of is the Decline and Fall of the British Empire?'[102] Taking this theme forward in her second Royal Commission on India, Nightingale similarly warned the British government that they would not 'hold' onto India if they didn't attend to the health of their soldiers who were stationed there.[103] 'Upon the British race alone the integrity of that empire at this moment appears to depend' she urged in 1858, reminding her readers in a tone filled with classic imperial fervour that '[t]he conquering race must retain possession'.[104]

Although Nightingale subsequently widened her gaze from British military health in India to broader rural social reform for Indian villagers, she continued to espouse certain imperialist assumptions. To Nightingale, health and social improvement required that Indians accept the education of the West. This was a supposition of backwardness that discursively infantilised the very people she wanted to help. Bearing this critical framing in mind, her 1878 battle cry for humanitarian intervention in India—'[h]ave we no voice for these voiceless millions?'—can be seen not only as progressive, but also as reinforcing an image of Indian victimhood, one that further casts the colonisers as the appropriate redeemers.[105] Time and again, Nightingale slipped into language that reinforced ideas that Indian civilisation was stagnating and needed active guidance.[106] For example, in 1894, she acknowledged that improvements could be made in the villages by the villagers themselves, but only 'if they can be *made* to understand the terrible results of neglect and the benefit of the most simple remedies'.[107] Similarly, Nightingale's later work to reform the Indian landholding system was organised around the central principle that it was an imperial responsibility to protect the rural poor who had no political voice.[108] She may have been keen to reform and improve the *zemindari* system in the 1870s, but this was notably to be achieved through an education in English history and literature.[109] Similarly, Nightingale stressed that Indian women needed to be educated into 'what makes home home'.[110] Furthermore, it was from English women they could best learn as they moved along their 'path of social progress'.[111] It is significant that when she conceived the scheme to introduce a nursing system into India, Nightingale recommended sending English-trained nurses to start the project off. Even though she stated her hope that Indian women would also eventually be able to participate in nurse training, she could not let go of the notion that, even once local nurses were appropriately trained, the matrons should always be European.[112]

Whether at home in the UK or abroad in empire, Nightingale's schemes were implemented and maintained through the strategies of discipline and surveillance that were the hallmarks of the imperial mindset.[113] Furthermore, as Britain amassed its territories of empire, the responsibility and configurations of what it meant to be British also gradually changed.[114] In this light, making empire 'home' can be configured as a profoundly political project, even if individual contributions to it, such as Nightingale's, were relatively Liberal in terms of some of her specific recommendations. Cook's description of Nightingale as 'Governess of the Governors of India' used a telling household metaphor in describing the importance of her role.[115] The model for reform in India was ultimately one that presumed Indian society needed to be brought into line with Western standards of behaviour, education, and culture. For all her liberalism, cultural sensitivity, bottom-up emphasis, and support of Indian independence, Nightingale was never able to totally break free of the notion that educated British citizens had a tacit right to offer tutelage over how India should be run.

The Limits of Bedroom Imperialism

Although it has been argued that in empire 'the isolation of rulers from ruled was integral to the colonial experience', Nightingale's bedroom imperialism had its obvious limits.[116] It might have become commonplace to live as a ruler in India in a separate European city enclave, or cool hill station, but advising over how such a vast imperial possession should be managed from a bedroom in dark, industrial London meant that—however much Nightingale read and studied the evidence or heard expert testimony—she never saw the problems that she hoped to address first-hand. Furthermore, the way she worked meant that she had an 'intense but flawed' reliance on favoured individuals.[117]

In fact, Nightingale felt that her whole India project was one of the chief failures of her career.[118] Despite her delivery of lengthy recommendations, amounting to many thousands of words, the India Office was slow and reluctant to follow her advice, and few tangible policy changes came about directly from her campaigns. Nightingale's Indian reform plans were unlikely to be very popular at Whitehall: not only were they potentially very expensive, they also threatened to expose the failures of imperial management. As Jharna Gourlay has noted, Nightingale therefore 'received little cooperation from the India Office' despite the 'verbal respect [that] was paid to her'.[119] Medical historians have also assessed Nightingale's involvement in India as limited. Her publication record on India may have been huge, but it was because of other prominent Western sanitary reformers working in situ in India, and work by reformers in the Indian Medical Service, that substantive changes were eventually made.[120] Being out of touch with the situation on the ground, Nightingale, as pointed out by Mridula Ramanna, increasingly 'made mistakes', particularly in terms of her presumptions over Indian needs.[121] Although the evidence is difficult to locate for Anglophone historians, some Indian vernacular newspapers have also been cited as criticising Nightingale's approach, particularly mentioning her lack of personal experience on the subcontinent as a fundamental drawback to her self-appointed role as sanitary advisor and social reformer. If she had been in India, one Hindi editorial bitterly commented, she would have seen that 'Indian villages are not the "happy villages" of England'.[122] Her internment as a perpetual (even if at times willing) housebound invalid made Nightingale's advice about India open to the criticism that it was too depersonalised and disconnected.

Nightingale's attitudes were both radical and conservative. She challenged class-bound imperatives that dictated that women should not work, yet the latter part of her life saw her permanently segregated from active professional life in her home. She supported greater opportunities for women, but, like many of her contemporaries, drew on gendered ideas of home and motherhood in order to realise these. She may have been unusual in the way she reformulated invalidism as prodigious productivity, yet in retrospect she did comparatively little to change the course of Indian sanitary history. Similarly, paradoxically, while she increasingly campaigned against imperialist attitudes and at times displayed remarkable sensitivity in terms of understanding the differences in contexts between India and England, she was nevertheless ultimately, to borrow Catherine Hall and Sonya Rose's term, 'at home with Empire' in crucial ideological ways.[123] Working from home, and hobbled by brucellosis, India stretched the reach of Nightingale's influence to its limit. It did offer her, however, a place for useful work to support and serve the British nation state. Nightingale received India papers at her South Street London home until she was well into her eighties. India represented a formidable forty-year involvement—the final home for Nightingale's remarkable professional energy.

Notes

1. The retrospective diagnosis of brucellosis was first made by D. A. B. Young, 'Florence Nightingale's Fever', *British Medical Journal* 311 (December 1995): 1697–1700. For evidence that Nightingale was still ill with brucellosis in the late 1880s, see Barbara Dossey, 'Florence Nightingale: Her Crimean Fever and Chronic Illness', *Journal of Holistic Nursing* 28, no. 1 (2010): 38–53.
2. Young, 'Nightingale's Fever', 1697.
3. An exception is Jharna Gourlay, *Florence Nightingale and the Health of the Raj* (Aldershot: Ashgate Publishing, 2003).
4. Nightingale, *Note on the Aboriginal Races of Australia: A Paper Read at the Annual Meeting of the National Association for the Promotion of Social Science, Held at York, September, 1864* (London: Emily Faithfull, 1865), https://wellcomecollection.org/works/y2jdtcxp (accessed 28 April 2020).
5. See Judith Godden, 'The Dream of Nursing the Empire', in *Notes on Nightingale: The Influence and Legacy of a Nursing Icon*, ed. Sioban Nelson and Anne Marie Rafferty (Ithaca: Cornell University Press, 2010), 55–75.
6. Christopher Hibbert, *The Great Mutiny: India 1857* (Harmondsworth: Penguin, 1978); Crispin Bates and Marina Carter, eds., *Mutiny at the Margins: New Perspectives on the Indian Uprising of 1857: Volumes I–VII: Documents of the Indian Uprising* (New Delhi: Sage, 2013–2017).
7. David Arnold, *Colonizing the Body: State Medicine and Epidemic Disease in Nineteenth Century India* (Berkeley: University of California Press, 1993); Mark Harrison, *Public Health in British India: Anglo-Indian Preventative Medicine, 1859–1914* (Cambridge: Cambridge University Press, 1994); Alison Bashford, *Imperial Hygiene: A Critical History of Colonialism, Nationalism and Public Health* (Basingstoke: Palgrave, 2004).
8. Nightingale, draft letter to the crown princess of Prussia, 22 October 1861, in *The Collected Works of Florence Nightingale, Volume 15: Florence Nightingale on Wars and the War Office*, ed. Lynn McDonald (Waterloo, ON: Wilfrid Laurier University Press, 2011), 621; Nightingale to Mary Carpenter, 3 August 1866, in *The Collected Works of Florence Nightingale, Volume 9: Florence Nightingale on Health in India*, ed. Gérard Vallée (Waterloo, ON: Wilfrid Laurier University Press, 2006) [hereafter *CW* 9], 562.
9. Mary Poovey, *Uneven Developments: The Ideological Work of Gender in Mid-Victorian England* (Chicago: University of Chicago Press, 1988), 198.

10. Nightingale to Sidney Herbert, 2 November 1857, BL Add Mss 43394 f194.
11. Parthenope Nightingale to Elizabeth Herbert, undated [August 1856], BL Add Mss 43396: ff50–52.
12. George Pickering, *Creative Malady: Illness in the Lives and Minds of Charles Darwin, Florence Nightingale, Mary Baker Eddy, Sigmund Freud, Marcel Proust, Elizabeth Barrett Browning* (Oxford: Oxford University Press, 1974).
13. Miriam Bailin, *The Sickroom in Victorian Fiction: The Art of Being Ill* (Cambridge: Cambridge University Press, 1994), 33; Millicent Garrett Fawcett, *Some Eminent Women of Our Times* (London: Macmillan, 1889), 76.
14. Lytton Strachey, *Eminent Victorians* (New York: Harcourt, Brace and World, 1918), 140; possibly drawing on Edward Cook, *The Life of Florence Nightingale*, vol. 1 (London: Macmillan, 1913), 148.
15. On Nightingale's opposition to idleness, see Jill Matus, 'The "Eastern Woman Question": Martineau and Nightingale Visit the Harem', *Nineteenth-Century Contexts* 21, no. 1 (1999): 63–87, 70–72.
16. On home being the place where 'one tends to feel more in control of things', see Michael Allen Fox, *Home: A Very Short Introduction* (Oxford: Oxford University Press, 2016), 115.
17. Pickering, *Creative Malady*, 127–128.
18. Mark Bostridge, *Florence Nightingale: The Woman and Her Legend* (London: Penguin, 2008), 329.
19. Bailin, *Sickroom*, 213. See also Rachel Ablow, 'Harriet Martineau and the Impersonality of Pain', *Victorian Studies* 56, no. 4 (2014): 675–697.
20. Nightingale, 'Cassandra', in *The Collected Works of Florence Nightingale, Volume 11: Florence Nightingale's Suggestions for Thought*, ed. Lynn McDonald (Waterloo, ON: Wilfrid Laurier University Press, 2008), 555–556.
21. Barbara Dossey, *Florence Nightingale: Mystic, Visionary, Healer* (Pennsylvania: Springhouse, 1999), 212–213.
22. Bostridge, *Florence Nightingale*, 325.
23. Ibid., emphasis added.
24. Ibid., 326.
25. Nightingale to Lady Canning, 16 September 1857, *CW* 9, 47.
26. Nightingale to John McNeill, 15 July 1857, *CW* 9, 49; Nightingale to James Pattison Walker, 3 January 1865, *CW* 9, 507.
27. Nightingale to James Pattison Walker, *CW* 9, 507.
28. Nightingale to Henry Bonham-Carter, 19 October 1888, *CW* 9, 9.
29. Gérard Vallée, 'Nightingale's Work on India', *CW* 9, 1–36, 27–28.
30. Chieko Ichikawa, 'Writing as a Female National and Imperial Responsibility: Florence Nightingale's Scheme for Social and Cultural Reforms in

England and India', *Victorian Literature and Culture* 39 (2011): 87–105, 101; Sandra Stotsky, 'Writing in a Political Context: The Value of Letters to Legislators', *Written Communication* 4 (1987): 394–410.
31. Liz Stanley, 'The Epistolarium: On Theorizing Letters and Correspondences', *Auto/Biography* 12 (2004): 201–235.
32. Judith Flanders, *The Victorian House* (London: Harper, 2003), xxvi.
33. Cook, *Life*, vol. 1, 324. On Nightingale's 1856 visit to Balmoral, see Lucy Worsley, *Queen Victoria: Daughter, Wife, Mother, Widow* (London: Hodder & Stoughton, 2018), 234–251.
34. Cook, *Life*, vol. 1, 497 for details of these residences.
35. Bostridge, *Florence Nightingale*, 303.
36. Deborah Anna Logan, ed. *Lives of Victorian Political Figures III, Volume 2: Florence Nightingale* (London: Pickering and Chatto, 2008), xiii; Dossey, *Florence Nightingale*, 194.
37. Note by Nightingale, undated [1857], in *The Collected Works of Florence Nightingale, Volume 5: Florence Nightingale on Society and Politics, Philosophy, Science, Education and Literature*, ed. Lynn McDonald (Waterloo, ON: Wilfrid Laurier University Press, 2003), 232.
38. Bostridge, *Florence Nightingale*, 408.
39. Lynn McDonald, 'Introduction' to *The Collected Works of Florence Nightingale, Volume 1: Florence Nightingale: An Introduction to her Life and Family*, ed. Lynn McDonald (Waterloo, ON: Wilfrid Laurier University Press, 2001) [hereafter *CW* 1], 1–84, 26.
40. Bostridge, *Florence Nightingale*, 410.
41. Ibid., 411.
42. Dossey, *Florence Nightingale*, 254.
43. Nightingale took Mai's departure in 1860 as an abandonment; it took her twenty years to be properly reconciled thereafter. Also worth mention is her cousin Henry Bonham-Carter, though he worked only on the Nightingale Fund/School. Dossey, *Florence Nightingale*, 244; on John Sutherland, 257.
44. Nightingale to Mary Jones, 7 February 1873, *CW* 1, 200.
45. Ibid.
46. Flanders, *Victorian House*, xxv.
47. Ida O'Malley, *Florence Nightingale 1820–1856: A Study of Her Life Down to the End of the Crimean War* (London: Butterworth, 1931), 345.
48. Cook, *Life*, vol. 1, 498–499.
49. Nightingale, Last Will and Codicils, *CW* 1, 852–861.
50. Flanders, *Victorian House*, 316–317.
51. Bostridge, *Florence Nightingale*, 411.
52. Nightingale, unsigned letter probably to William Nightingale, 9 January 1870, *CW* 1, 735–736.

53. Bostridge, *Florence Nightingale*, 411.
54. Ibid., 412; Mr Bismark (*sic*), Mufti, Tit, Tom, and Topsy are mentioned frequently by name. For example: Nightingale to Mary Mohl, 18 April 1870, *CW* 1, 759; see also McDonald's description *CW* 1, 755.
55. Joseph Chamberlain, 'A Young Nation', speech given at the Imperial Institute, 11 November 1895, cited in J. L. Garvin, *The Life of Joseph Chamberlain*, vol. 2 (London: Macmillan, 1933–35), 27.
56. Best epitomised by Liberty's department store, which sold Japanese and Eastern luxury goods from 1875 and was joined in the 1880s by the 'Eastern Bazaar' which showcased Indian artisanal fabric, ornaments, and *objets d'art*. See James Laver, *The Liberty Story* (London: Liberty, 1959).
57. Godden, 'Nursing the Empire', 55–75, 60.
58. Kaori Nagai, 'Florence Nightingale and the Irish Uncanny', *Feminist Review* 77 (2004): 26–45, 28.
59. Frank Prochaska, *Women and Philanthropy in Nineteenth-Century England* (Oxford: Clarendon Press, 1980), 41–42; Jane Rendall, 'The Condition of Women, Women's Writing and the Empire in Nineteenth Century Britain', in *At Home with Empire: Metropolitan Culture and the Imperial World*, ed. Catherine Hall and Sonya Rose (Cambridge: Cambridge University Press, 2006), 101–121, 121.
60. R. M. George, *The Politics of Home: Postcolonial Relocations and Twentieth-Century Fiction* (Cambridge: Cambridge University Press, 1996), 36.
61. Marie Ruiz, *British Female Emigration Societies and the New World, 1860–1914* (London: Palgrave, 2017).
62. Vallée, 'Nightingale's Work on India', *CW* 9, 15.
63. Nightingale, *Observations on the Evidence Contained in the Stational Reports Submitted to her by the Royal Commission on the Sanitary State of the Army in India* (London: E. Stanford, 1863), *CW* 9, 130–183.
64. Nightingale, notes from a meeting with Edward Stanhope (Secretary of State for War), 1 November 1888, in *The Collected Works of Florence Nightingale, Volume 10: Florence Nightingale on Social Change in India*, ed. Gérard Vallée (Waterloo, ON: Wilfrid Laurier University Press, 2007) [hereafter *CW* 10], 388.
65. Nightingale, 'How People May Live and Not Die in India', *Transactions of the National Association for the Promotion of Social Science* (London: Longmans, 1864), *CW* 9, 184–194.
66. Nightingale, 'Suggestions on a System of Nursing for Hospitals in India, a Letter to the Secretary of the Sanitary Commission for Bengal', 24 February 1865, *CW* 9, 950–955.
67. Nightingale to G. B. Malleson, 26 September 1867, *CW* 9, 973–974.
68. Lynn McDonald, *Florence Nightingale: Nursing and Healthcare Today* (New York: Springer, 2018), 50.

69. See the 'Epilogue' to *CW* 10, 890–902, here 893. For more on the development of public health in India, see Harrison, *Public Health*.
70. Nightingale, 'Life or Death in India: A Paper Read at the Meeting of the National Association for the Promotion of Social Science, Norwich, 1873', *CW* 9, 710–750, 723.
71. Nightingale, 'Our Indian Stewardship', *The Nineteenth Century* 114, no. 18 (1883): 329–338, *CW* 10, 819–830; Logan, ed. *Lives*, xv.
72. Mohandas K. Gandhi, 'No. 80, Florence Nightingale', in *The Collected Works of Mahatma Gandhi*, vol. 5, 1905–1906 (New Delhi: Ministry of Information and Broadcasting, 1994), 61–62, cited in *CW* 10, 848.
73. Onni Gust, '"The Perilous Territory of Not Belonging": Exile and Empire in Sir James Mackintosh's Letters from Early Nineteenth-Century Bombay', *History Workshop Journal* 86 (2018): 22–43.
74. Original underlining. Nightingale to 'Colonists of South Australia', 28 January 1858, Claydon N307, Wellcome 8997/62.
75. The unifying idea of a 'Greater Britain' was particularly forwarded within Britain's settler colonies, as a model for both political and cultural unity. See Duncan Bell, *The Idea of Greater Britain: Empire and the Future of World Order, 1860–1900* (Princeton: Princeton University Press, 2011).
76. Tony Ballantyne and Antoinette Burton, *Empires and the Reach of the Global, 1870–1945* (Cambridge, MA: Harvard University Press, 2012), 24.
77. Ibid., 80–90.
78. Fredrick Cooper and Ann Laura Stoler, eds. *Tensions of Empire: Colonial Cultures in a Bourgeois World* (Berkeley: University of California Press, 1997).
79. Anne McClintock, *Imperial Leather: Race, Gender and Sexuality in the Colonial Context* (London: Routledge, 1995).
80. Ruth Rogaski, *Hygienic Modernity: Meanings of Health and Disease in Treaty-Port China* (Berkeley: University of California Press, 2004), 3.
81. Mary Poovey, *Making a Social Body: British Cultural Formation, 1830–1864* (Chicago: University of Chicago Press, 1995), 115–131; J. H. L. Cumpson [Australian public health bureaucrat] quoted in Bashford, *Imperial Hygiene*, 1.
82. Nightingale to Sir John Lawrence, 26 September 1864, *CW* 9, 212.
83. Poovey, *Uneven Developments*, 166.
84. Nightingale, 'Health Lectures for Indian Villages', *India* (October 1893): 305–306, *CW* 10, 375–377; 'Health Missioners for Rural India', *India* (December 1896): 359–360, *CW* 10, 388–392.
85. Nightingale, *Stational Reports*, *CW* 9, 144. See also Nightingale, *Suggestions in Regard to Sanitary Works Required for Improving Indian Stations*, prepared by the Barrack and Hospital Improvement Commission, 15 July 1864, Blue Book (London: Eyre and Spottiswoode, 1864), *CW* 9, 302–730; 340–342.

86. Nightingale, 'Rural Hygiene', *Official Report of the Central Conference of Women Workers* (1894), 46–60, in *The Collected Works of Florence Nightingale, Volume 6: Florence Nightingale on Public Health Care*, ed. Lynn McDonald (Waterloo, ON: Wilfrid Laurier University Press, 2004) [hereafter *CW* 6], 607.
87. Nightingale to James Pattison Walker, 3 January 1865, *CW* 9, 507.
88. Nightingale frequently used this term to refer to Indians, e.g. in an 1877 letter to the editor regarding the Madras famine, *The Illustrated London News*, 7 July 1877, 22–23, *CW* 9, 761.
89. Vallée, 'Nightingale's Work on India', *CW* 9, 10; see also Gourlay, *Health of the Raj*, 268.
90. Nightingale, 'The People of India', *Nineteenth Century* 4 (1878): 193–221, 193. On the various versions of this text see Vallée, '"The Zemindar, the Sun and the Watering Pot," 1874', *CW* 10, 401–404.
91. Nightingale, 'The United Empire and the Indian Peasant', *Journal of the National Indian Association* 90 (July 1878): 232–245, *CW* 10, 487–500; 'The Dumb Shall Speak and the Deaf Shall Hear; or the Ryot, the Zemindar and Government', *Journal of the East India Association* 15 (1883): 163–210, *CW* 10, 548–598.
92. Nightingale, 'The United Empire', *CW* 10, 489–490.
93. Ibid., 497.
94. Note by Nightingale, 14 February 1853, *CW* 10, 797.
95. Nightingale, 'Health Lectures', *CW* 10, 375.
96. Ibid., 376.
97. Nightingale, 'Health Missioners', *CW* 10, 391.
98. Nightingale, 'Rural Hygiene', *CW* 6, 607.
99. Ichikawa, 'Writing as a Female National', 89.
100. Poovey, *Uneven Developments*, 191.
101. Gourlay, *Health of the Raj*; Vallée, 'Indian Society and Culture in Transition', *CW* 10, 807.
102. Nightingale to Lieutenant Colonel Lefroy, 6 March 1856, in *The Collected Works of Florence Nightingale, Volume 14: The Crimean War*, ed. Lynn McDonald (Waterloo, ON: Wilfrid Laurier University Press, 2010) [hereafter *CW* 14], 343–345, 344.
103. Mridula Ramanna, 'Florence Nightingale and the Bombay Presidency', *Social Scientist* 30, no. 9 (2002): 31–46; 43.
104. Nightingale, *Notes on the Health of the British Army* (1858), in *CW* 14, 578–888, 883.
105. Nightingale, 'People of India', 193.
106. Mrinalini Sinha, *Colonial Masculinity: The 'Manly Englishman' and the 'Effeminate Bengali' in the Later Nineteenth Century* (Manchester: Manchester University Press, 1995).

107. Emphasis added. Nightingale, *Village Sanitation in India* (London: Spottiswoode and Co., 1894), *CW* 10, 380–384, 382.
108. Nightingale, 'People of India', 193.
109. Nightingale, 'Can We Educate Education in India to Educate *Men*?' (Part 2), *Journal of the National Indian Association* 105 (1879): 478–491, 480–481; *CW* 10, 644–654, 647.
110. Nightingale, 'Can We Educate Education in India to Educate *Men* and *Women*?', *Journal of the National Indian Association* 106 (1879): 527–558, 541, *CW* 10, 654–676.
111. Nightingale, 'Introduction', in Dayārāma Gidumal, *Behramji M. Malabari: A Biographical Sketch* (London: Fisher Unwin, 1892), v–viii, viii.
112. Nightingale, 'Suggestions on a System of Nursing for Hospitals in India' (1865), *CW* 9, 948–980.
113. Ichikawa, 'Writing as a Female National', 91. See also Cynthia Hammond, 'Reforming Architecture, Defending Empire: Florence Nightingale and the Pavilion Hospital', in *(Un)Healthy Interiors: Contestations at the Intersection of Public Health and Private Space*, ed. Aran S. MacKinnon and Jonathan D. Ablard (Carrollton, GA: University of West Georgia, 2005), 1–24, 11.
114. Daniel Gorman, *Imperial Citizenship: Empire and the Question of Belonging* (Manchester: Manchester University Press, 2007).
115. Edward Cook, *The Life of Florence Nightingale*, vol. 2 (London: Macmillan, 1913), 169.
116. Ranjit Guha, 'Not at Home in Empire', *Critical Inquiry* 23, no. 3 (1997): 482–493, 484.
117. Godden, 'Nursing the Empire', 75.
118. Certainly, this is how her biographer Cook positioned her feelings. See the review in *The Spectator*, 15 November 1913, 823–824.
119. Gourlay, *Health of the Raj*, 192.
120. Arnold, *Colonizing the Body*, 98; Harrison, *Public Health*, 65.
121. Ramanna, 'Nightingale and the Bombay Presidency', 32.
122. *Kaiser-I-Hind*, Bombay, 27 March 1887, translated in 'Miss Nightingale on Village Sanitation in India', *Voice of India* (April 1887): 188–189, cited in Ichikawa, 'Writing as a Female National', 103.
123. Hall and Rose, eds. *At Home with Empire*; see also Cooper and Stoler, eds. *Tensions of Empire*.

CHAPTER 8

Spiritual Home

Faith and divinity have long had a complex relationship with the home. Over many centuries domestic spaces such as private chapels, bedsides, or hearths have, in Thomas Barrie's words, 'house[d] the divine'.[1] In nineteenth-century Britain, the relationship between religious observance and the home came under tension, as middle-class society struggled to reconcile piety with a growing consumerist predilection for acquiring material possessions. The result, as Deborah Cohen has stressed, was that domestic consumerism itself was reframed as semi-religious: the British middle classes redefined 'consumption as a moral act, and the home as a foretaste of the heaven to come'.[2] The Victorians extended their idealisation of home and domesticity onto their understanding of heaven—and vice versa, imbuing their homes with sacred qualities.

Florence Nightingale's own Christian faith was central to her ideas of work, as well as to her personal relationships. Like so much else in her life, her faith was conditioned by the material realities and ideological constructions of home. Particularly in her post-Crimea housebound years, Nightingale's home was the principal site of her spiritual expression, prayer, and religious ritual. In addition, faith was an important element in her public reputation and posthumous image: bound up in her celebrity in her lifetime and in the way she was subsequently memorialised—or, in effect, canonised—beyond her death.

Nightingale was an itinerant, restless, heterodox Christian, with a faith that was always strong and often intense. While her mother's and father's

P. Crawford et al., *Florence Nightingale at Home*,
https://doi.org/10.1007/978-3-030-46534-6_8

backgrounds were influenced by Unitarian nonconformism, by the time of Nightingale's birth her immediate family were Church of England.[3] As a young woman, she attended and financially supported different chapels and churches in her quest to find a home for her developing faith, a journey partly conditioned by her own changing home environment.[4] While sometimes probing and questioning certain tenets of religion during these early years, her faith was, to use Lynn McDonald's words, 'the great constant and centre of her being'.[5] She felt called to serve God with such spiritual conviction that she heard voices, presumed divine, on at least four occasions, and experienced a number of other religious visions and what she called 'impressions'.[6] During her spells at Lea Hurst, she attended Methodist and Baptist chapels—both nonconformist faiths being popular in Derbyshire at the time—besides Church of England services.[7] In later life she remained, broadly speaking, a liberal Anglican with some Methodist sympathies, but explored ecumenism with visits to both Protestant and Roman Catholic religious communities: the former at Kaiserswerth (Chapter 5) and the latter when she visited the Sisters of Charity in France, Italy, Ireland, and Egypt in her 1848–1853 period of exploration and preparation for an active career. Nightingale was tempted to convert to the Catholic Church, which, following its growth in influence in Britain in the middle of the century, promised appealing opportunities for unmarried religious women to pursue a vocation. In 1852 she undertook a revealing correspondence with the (later Cardinal) Henry Manning, a prominent Catholic convert who sought to convert Nightingale also. Becoming Catholic, Nightingale could see, would make it easier for her to undertake nursing or other humanitarian work. 'What a home the Catholic Church would be to me', she wrote, sensing that she would find in it 'my work, already laid out for me, instead of seeking it to and fro and finding none: my home, sympathy, human and divine'.[8] Some women in Nightingale's circle did indeed convert. Among them, notably, was her friend Elizabeth Herbert, who helped her co-ordinate nursing recruits for the Crimean War; as well as Mary Stanley, who led the second party of war nurses in December 1854; and, later, Angélique Lucille Pringle, the second superintendent of the Nightingale School.

Herbert described her conversion as a homecoming in her book *How I Came Home* (1894). Yet despite her own clear attraction to Catholicism's offer of a spiritual home, Nightingale held back, apparently due to theological reservations. The problem, she told Manning, was that the Catholic Church 'insists peremptorily upon my believing what I cannot believe'.[9] As much as Nightingale wished for the Anglican Church to reform and to offer the kinds of opportunity to women that the Catholic Church provided, she refused to abandon it.[10] Nightingale remained keen, however, to explore many different perspectives on faith and spirituality, thinking and writing expansively about everything from Hinduism, Brahmanism, Buddhism, Judaism, and Pantheism to ancient Egyptian and Greek forms of worship.[11] On her 1849–1850 Nile trip she felt a particular connection to ancient Egyptian religious sites, writing at the decaying monuments at Karnak that 'their God was my God' and explaining how the rock temples at Abu Simbel made her feel 'more *at home*, perhaps than in any place of worship I was ever in'.[12] Relatively unusually for the period, but in keeping with her liberalism, Nightingale's comparative views on non-Christian religions were well-informed and broadly tolerant, once in a personal note calling for 'the Hindu, the Buddhist, the Christian [to] each live in his God's sight, doing His work rightly'.[13]

As a prodigious worker and frequent critic of idleness, Nightingale's views on religion were unsurprisingly bound up with her ideas on work. Her pragmatic conclusion was that Christians should seek to do God's work in the world rather than passively wait for God to solve human problems. In *Suggestions for Thought* (1860), Nightingale challenged references to a cholera outbreak as somehow being part of God's plan, expressing surprise at those who prayed to be 'delivered from "plague, pestilence and famine," when all the common sewers ran into the Thames, and fevers haunted undrained land, and the districts which cholera would visit could be pointed out'. Nightingale's religion was one in which God wanted humans to use modern scientific understanding to improve their lot. Cholera and other diseases were not a divine punishment for sin; tackling their causes required not prayer or inner purification but active work in the world. 'I thought that cholera came that we might remove these causes, not pray that God would remove the cholera', she reflected.[14] Mysticism, for her, was an insufficient response to the challenge of living a Christian life: the task was to emulate Christ's material impact on the world. 'It is very easy to be religious, if religion is only the getting up to God (mysticism)', she wrote. 'But it is very difficult to be religious in the sense of incarnating Him upon earth'.[15]

Nightingale also unavoidably approached religion as a woman and emphatically as a woman seeking the work and refuge that were closely connected in her mind. Visiting a Cairo mosque in 1850, she wrote that:

> It is so pleasant to see a place where any man may go for a moment's quiet, and there is none to find fault with him nor make him afraid. Here the homeless finds a home, the weary repose, the busy leisure; if I could have said where any *woman* may go for an hour's rest, to me the feeling would have been perfect, perfect at least compared with the streets of London and Edinburgh where there is not a spot on earth a poor woman may call her own to find repose in.[16]

This passage shows the appeal of the idea of a spiritual shelter, as well as Nightingale's acute perception that gender inequalities extended to the spiritual domain. It further demonstrates the tendency 'to compare and contrast her own situation to that of women elsewhere' that Jill Matus has observed in Nightingale's writing on Egypt.[17]

Away from the divine appeal of churches, chapels, temples, and mosques, Nightingale considered one's personal home as the locus of faith. As shown below, she endeavoured at various times to create a 'household of faith' that attended to the private, spiritual needs of residents, whether relatives, companions, or employees.[18] On an even smaller scale, she conceived of her own body as a temple that housed, or had the potential to house, divinity. In 1892 she hoped that an 'indwelling God' might find a way of entering 'into my heart and dwelling and drive me out'.[19] It was in this way that Nightingale pursued a more intimate devotional life in the domestic setting, secure in her belief in a loving and certainly not wrathful God.[20] Her vision was of a merciful and compassionate God fit for any home: a companionable and homely kind of deity. Nightingale understood and invested in the metaphysical dimension of the home and came to embody the wish for what Rowan Moore has described as a 'shelter both physical and spiritual'.[21]

Households of Faith

Any house can serve as a place of worship and faith as well as a domestic dwelling. Homes can act as sanctuaries for thought and contemplation, in which religious experience can occur in ways that are comparable to more formal institutions of faith.[22] Nightingale, certainly, did not believe that God resided only in consecrated institutions. On Good Friday 1850, she recorded that she 'stayed at home … knowing that I did not go to church to seek God nor expect to find Him there'.[23] The alternative focus on informal or 'low' kinds of worship that do not require a formal liturgy, church building, or traditional leader figure was a feature of early Methodism, particularly Primitive Methodism, and its principle of the 'priesthood of all believers'.[24] In addition, Nightingale would have been aware that her Unitarian ancestors were obliged between 1697 and 1813 to worship principally in private houses, owing to legal restrictions on the preaching of non-Trinitarian doctrine. Indeed, it was her own grandfather, the Whig MP William Smith, who in 1812–1813 carried important measures though Parliament to extend religious liberty and end this ban.[25]

In Nightingale's 1877 diary (which, unusually, survived her significant culls of personal papers), she referred several times to maintaining a household of faith. This idea included overseeing the spiritual lives of her domestic employees, giving them religious lessons, and supporting their progress through important Christian rites. 'O God, a household of faith: Fanny, Annie, Polly', Nightingale wrote in an entry marked 17 June, referring to, respectively, her maid, her cook, and her maid's sister.[26] The other diary entries of spring 1877 reinforce this impression of Nightingale's religious, familial-like responsibility towards her employees, whether making arrangements for their confirmation or praying for their first communion.[27] Nightingale also gave Bibles as gifts to family members, friends, and employees such as Alice Munday, who was gifted a volume in 1875 inscribed with Nightingale's 'affectionate good wishes and earnest prayers for her best happiness'.[28] The idea of securing a household of faith occupied her into her late life. 'How can I make "Thy kingdom" in this house?', she asked God in an 1893 journal entry, repeating her questioning of the sacred within the domestic in 1899: 'Is *this God's house*? Is this room God's house?'[29] What emerges from these spiritual dialogues is the intensity of Nightingale's religious feeling, albeit regularly laced with some accompanying doubt.

In the 1877 diary Nightingale also describes her experience of personal prayer, noting how, when '[d]own upon my knees in armchair', she would '[v]oiceless, pray [to] God about Indian work'.[30] This religious practice transformed the mundane into the godly, the routine into the exceptional. 'Prayer is when I get up in the morning not to get up because it will be unusual if I do not, because I shall be too late for breakfast or too late for my day's toil', she explained in an earlier sermon, 'but to get up to do God's work'.[31] When faith was brought into the everyday in this way, even housework became a sacred act:

> How am I to tell what is God's will for the little things of every day in order that I may obey it? God is always speaking in the circumstances of our everyday life if we will but listen and honestly ask Him, 'Lord what wilt Thou have me to do?' The little housemaid, who modestly said in answer to a question asked her in class that she thought she had grace, when asked why she thought so, truly answered 'Because I sweep under the mats.' What is grace? So the old hymn says: Who sweeps a room as by His laws / Makes that and the action fine.[32]

Here, God's plan is manifest everywhere, even in the most straightforward actions in the home. In terms that recall the dutiful sanitary commandments of *Notes on Nursing* (Chapter 4), Nightingale spots faith lurking under the rug.

Confined to her house, Nightingale prayed from the privacy of her armchair rather than the pew, or from her bed or bedchamber rather than at an altar. Her views on prayer were not straightforward or uncritical. She doubted the power of prayer in petitioning direct, simplistic answers from God, and, during her period of profound religious questioning in the 1840s and 1850s, even subjected her prayers to an empirical test: 'In private prayer I wrote down what I asked for, specified the time by which I prayed that it might come, continued in prayer for it, and looked to see whether it came. It never did'.[33] Preferring to pursue God's work in the world rather than wait on a divine intervention that would most likely not come, Nightingale considered prayer as something nebulous, beyond satisfactory description: '[w]e cannot dogmatize on this highest intercourse. There can be no "form of prayer" which will be the voice from all hearts', she wrote.[34]

Beyond her prayerful devotion and spiritual oversight of employees, Nightingale also took on the role of godmother to many children beyond her immediate household, including those of Pastor Fliedner, Sir John McNeill, and Samuel Gridley Howe.[35] She showed interest in her godchildren, sending them religious and other kinds of gifts such as Bibles, encouraging letters, and sometimes money.[36] The satisfaction that she drew from these relationships is evident in an 1881 letter to her step-nephew Edmund Verney on the birth of his son Harry. 'Tell my little Ruth [Harry's older sister] to send her godmother a detailed account of the young hero whose protector and guardian she now is', Nightingale wrote.[37] Her genuine affection in this role is clear by her signing later letters to Harry and Margaret Verney, another step-nephew and step-niece, 'ever your loving Aunt Florence'.[38]

The Bible was core home reading throughout Nightingale's life, alongside a variety of devotional and spiritual literature. Her copious and sophisticated biblical annotations frequently allude to the spiritual life in terms of home. For Nightingale, God resided in the roomy universe, '[b]ut as Almighty God cannot but perceive and know everything in which he resides, infinite space gives room to infinite knowledge and is as it were an organ of omniscience'. She conceived of the body as housing the soul and accounted for spiritual interiority in architectural terms as 'our inner chambers' and 'small secret recesses'.[39] She advocated inviting God into the home: 'If we admit Him He will dwell in our hearts, if we consecrate our houses to Him He will there manifest His presence'.[40]

This earnest devotional reading is represented in a portrait by her cousin Hilary Bonham-Carter from 1854 in which Nightingale appears as a seated, angelic figure in white gazing at a book, a crucifix hanging on a ribbon around her neck suggesting that the reading is spiritual or devotional (Fig. 8.1). It is interesting to place this image in the context of the letters that author Elizabeth Gaskell wrote from Lea Hurst in October 1854 recounting stories about Nightingale gathered from family recollection and anecdotes. In these letters Gaskell repeatedly refers to Nightingale as semi-divine, 'like a saint', standing 'half-way between God and His creatures', and engaged on a 'visible march to Heaven'.[41] When considered together, Gaskell's letters and the Bonham-Carter image suggest that at the time of her departure for the Crimean War Nightingale's family circle was at least half-seriously entertaining the idea that she might indeed be divinely inspired. This may have helped them to overcome any lingering doubts about the respectability of Nightingale's career choice (Chapter 3).

Upon her return from the Crimean War, Benjamin Jowett, Regius Professor of Greek at Balliol College, Oxford, entered Nightingale's household of faith. After first approaching him in 1860 to solicit a response to *Suggestions for Thought*, Jowett subsequently became Nightingale's regular correspondent, long-time friend, and, purportedly, suitor.[42] He visited her regularly in London for discussions on spirituality or Plato's philosophy, and occasionally, being an ordained priest, provided the sacrament of communion to Nightingale at her home. This tradition of 'infrequent, semi-public form of worship' began in 1862 at Nightingale's request and subsequently became a mainstay of their friendship until

Fig. 8.1 Drawing of Nightingale reading a book, with a crucifix on a ribbon necklace. Joanna Hilary Bonham-Carter, 1854, reproduced in Sarah Tooley, *Life of Florence Nightingale* (London: Bousfield, 1904), 48

Jowett's death in 1893.[43] These rituals were one of the few occasions that Nightingale actively asked her parents and sister into her London home. 'I invite you and Papa', Nightingale wrote to her mother in June 1863, explaining that she had 'asked Mr Jowett (who will not be in London again for two or three months) to give me the sacrament next Sunday, 28th, at 3.00'.[44] She sent a similar invitation to her sister in February 1865 .[45]

These home sacraments were necessitated by Nightingale's illness. Her inability to worship at church was, she explained in 1870, 'one of [my] greatest afflictions', four years later bemoaning the extensive period that had passed since she had last stepped inside a Christian place of worship.[46] Sometimes, Nightingale's poor physical health undermined even her home-based prayers with Jowett, since the obligation to converse afterwards left her excessively tired. '[T]he fatigue to me of taking the sacrament is so great', she wrote to her mother in 1870, suggesting that she 'take Mr Jowett back to his house after the sacrament'.[47] On other occasions, however, Nightingale would delight in Jowett's presence and social eminence, especially after he became Vice-Chancellor of the University of Oxford in 1882.[48] Her homebound friendship with Jowett certainly seems to have assuaged her sense of spiritual isolation from those with less urgent religious convictions. Writing to Jowett in 1866, Nightingale described herself as a castaway in a largely faithless London: 'I must be Robinson Crusoe come to life again as a female – I think that, in London, out of a few special coteries, no one cares about "the great truths of religion", which generally means the very little truths or even the non-truths of religion'.[49] It seemed that Nightingale, the castaway, had found a loyal Man Friday in Jowett.

Nightingale and Jowett shared a spiritual and political outlook and worked well together. Nightingale drafted entire sermons for Jowett to preach, as in February 1869, when Jowett informed Nightingale of his delight 'to hear that you have written a sermon (as pleased as Phaedrus was when Socrates promised to make a speech). It shall certainly be preached (but you must not object to my taking out the oaths) and afterwards published. Let me have it without delay'.[50] Although it is unclear if in practice Jowett preached such contributions, he openly acknowledged her broader suggestions in 1871, writing that 'I shall take the text which you suggest [Luke 18.8] and endeavour to work it out'.[51] Nightingale was aware of the challenges inherent in writing and performing lively sermons, writing of their 'deadness' as a form when poorly written or delivered.[52] Yet, despite these misgivings, she was clearly enthused about authoring such texts, in 1869 proudly enclosing a package likely addressed to her father in an attractive presentation box with the words: 'I have sent a *sermon* of mine!'[53]

Nightingale sought spiritual influence and leadership by writing and sharing sermons with Jowett and others. As such, we can see her sermon writing as a kind of bridge between her sacred domesticity and more formal institutions of religion. Yet Nightingale's sermons, like her wider religious views, were unconventional, and sometimes at odds with mainstream Anglican beliefs. For example, she did not believe in the historicity of the resurrection story and, in one of her writings, disavowed the idea of the literal virgin birth.[54] While questioning the ultimate authenticity of such stories, Nightingale nevertheless grasped their symbolic power. In one intriguing reference to the '*idea* of the "virgin mother"', for instance, Nightingale reflected on Mary's example as an unmarried, 'motherly' spiritual guide, standing as 'the goodness of God':

> Though it is probable that the Virgin never lived at all, at least (or certainly) not as she is represented, yet there is a deeper truth in those to whom she stands as the goodness of God, and who find their best assurance of God being more than father, more than mother, to us in that beautiful fable than there is in those who call her by I know not what disagreeable words. Also: there can be no doubt, for all history, all society shows it us that there is a profound truth in the *idea* of the 'Virgin mother' – since it is *not* people's own fathers and mothers who influence them most or most generally for good.[55]

The passage avows the symbolic importance of the family and shows Nightingale attuned to the influence of religious figures upon everyday life. Her various efforts to establish and develop professional communities similarly drew upon the power of visionary, family-like, and sometimes sacred models to motivate work and vocation (Chapter 5). In the early stages of the Nightingale School and Nightingale Home, for example, probationers were exposed to the symbols of faith in voluntary Bible classes provided by the Home Sister, in addition to daily routines featuring prayers and religious themes. Later, Nightingale took up a clearer position as a role model to the nursing trainees herself by offering spiritual guidance and mentoring. By 1873 she was referring to her recruits as 'my children' and in her regular addresses referred to nursing as a calling from God.[56] 'What higher "calling" can we have than nursing?' she asked rhetorically in 1872.[57] A picture of Nightingale with her students during a visit to Claydon House underlines her status as a spiritual leader: the probationers are shown turning towards Nightingale as devotees, their seating arrangement suggesting a religious-like obedience towards the (by that time) legendary nurse (Fig. 8.2). Whether arranged by the photographer or at Nightingale's behest, the image captures the idea of Nightingale as influential virgin mother—a kind of lay abbess, or mother-chief, as she described herself to her trainees in 1900.[58] Just as God was to be discovered by conscientious work in the home, his sacred presence was here invoked through a commitment to worldly vocation. Nightingale's 1873 address to the nurses expressed the hope that '*life* and *work* among the sick' would 'become a prayer. For prayer is communion *or co*-operation with God, the expression of a *life* among his poor and sick and erring ones'.[59] A house of work and a household of faith were for Nightingale ultimately one and the same thing.

Fig. 8.2 Photograph of Nightingale and Sir Harry Verney with Nightingale School Probationer Nurses during an 1886 visit to Claydon House, Buckinghamshire, Wellcome L0010473

Heavenly Homes and House Tombs

If houses were sites of faith and homes for God, a Christian's ultimate home would lie in God's house: namely heaven. This idea became widespread in the Victorian approach to death, which increasingly stressed the domestic qualities of heavenly reward over the deterrent threat of hellfire punishment. In Boyd Hilton's words, '[h]ell departed into metaphor, while [h]eaven retained its felicitousness [and] became domesticated'.[60] As Pat Jalland has written, a Victorian death became 'ideally a family event interpreted in terms of a shared Christianity, with the assurance of family reunion in heaven'.[61] Death could therefore be thought of as simply a return home, a kind of homecoming to the original heavenly home of the lost Eden story in Genesis.[62] Jalland locates this sentimentality in such places as *Heaven our Home* (1861) by the Presbyterian minister William Branks, which featured the following passage:

> You have taken the last look of the home you are about to leave, and of the well-known, dear, but sorrowful faces that are around you … *How comforting to you, in your present circumstances, is the knowledge that heaven is your home!* Death to you will thus merely take you out of *one* home, that it may usher you into *another*… Around the very *word* 'home' what holy and sacred associations cluster and hang![63]

Nightingale's writings on theology, spirituality, and religious practice reflected this homely view of heaven. She frequently mused about death as passing through a door or gateway to God's 'dwellingplace'.[64] In 1869, she wrote in her journal that 'to pass through the valley of the shadow of death is the way home … my heart and soul have often pined for their home. Did I know where or what that home was?'[65] In an 1889 letter to her cousin Shore Smith on the death of his mother, Nightingale drew again on this analogy of a final journey home, stating that '[s]he went home to that home where she will be no stranger at 1.00 this morning, went to her God, after Whom she had longed'.[66] Similarly, Nightingale informed a Derbyshire carpenter of her own mother's death in 1880 by writing simply, 'my dear mother is gone home'.[67]

The seriousness with which the Victorians treated death and mourning is well known, epitomised by the Queen's decades-long mourning for Prince Albert. Nightingale observed mourning conventions carefully, as the black borders which frequently adorn her letters attest, and she marked the anniversaries of the deaths of key figures in her life, especially Sidney Herbert. She was diligent and thoughtful in writing letters of condolence to those who had lost loved ones, especially in the Crimean War when she provided dying soldiers in her care with the opportunity to communicate with their families back home (private correspondence that sometimes then found its way into the pages of the national and the regional newspapers).[68] In one instance, Nightingale comforted a mother on the death of her chaplain son at Scutari Hospital in 1855, writing: 'I cannot tell you the feeling of deep sympathy, with which I beg to enclose a lock of your poor son's hair. You will hear from others than me of his death and of your loss; I will only tell you of your gain. His last thought was of you'.[69] As McDonald points out, these letters of condolence use metaphors of heaven as home as a means of comforting the grieving, reassuring them that God 'knew what He was doing in taking them "home"'.[70]

Most Victorians did not pass away somewhere as distant as a battlefield, or even a hospital, but instead in the home itself. Death was therefore conceived not only as a symbolic form of homecoming, but also as a natural domestic event integral to family life. As such, the preparation for, and supervision of, death itself was subsumed among the range of domestic duties considered to be women's work. Nursing relatives during terminal care, laying out of the dead: these were tasks that fell to wives, mothers, and daughters—or occasionally, in richer families, female servants or nurses.[71] Nightingale performed such a role on various occasions in her personal and work life—for example, devoting eleven days of deathbed care to her grandmother Mary Shore at her home in Tapton, near Sheffield in 1853, prior to the death itself on 25 March. Her updates to her family during this vigil were full of poetic metaphors: 'The full moon shone on the waste of snow last night, as the face grew beautiful in the light of death and young in the hope of life. I almost wish you could have seen her as she is now'.[72] Nightingale treated her grandmother's bedsores with silver nitrate, informing her father that she had made his mother 'comfortable for death'.[73] In the 1870s, she combined managing her own poor health with attending to the sick around her, especially her mother. Her approach to end-of-life care seems to have been characteristically practical. In later life, unable to appear at the bedside of dying friends and relatives, Nightingale would instead send provisions, once arranging for a nun that she knew to receive fresh eggs, jelly, beef juice, port, and champagne on her deathbed.[74] At times, her own bed became a source of funereal prefiguration, and the irony of her bed serving as both her workplace and place of confinement was not lost on her. In 1877 she asked of God, 'how canst Thou take one as Thy hired servant, who is bedridden and unable?'[75] She had many premature forebodings of her own mortality, not least at the height of her illness in the late 1850s and early 1860s, yet, defying both her own and her doctors' expectations, survived into old age.

In her last will, drafted in the 1890s, Nightingale sought to balance family privacy with an awareness of herself as a kind of public property, clearly recognising that her name carried with it a national legacy and held particular responsibilities to successive generations of nurses and soldiers. She bequeathed most of her estate to various branches of her extended family, but also made specific provision for gifts or financial donations to several people with whom she had closely worked. These included the engineer Douglas Galton, the philanthropist William Rathbone, J. J. Frederick from the Army Sanitary Commission, and the daughters of the statistician William Farr, as well as John Croft and Mary Crossland, the instructor and Home Sister at the Nightingale School.[76] In tacit acknowledgement of her own celebrity, Nightingale requested that the jewels and medals she had received from Queen Victoria at the time of the Crimean War be displayed 'at some place where soldiers may see them' (they are now at the National Army Museum).[77] She additionally requested that her collection of paintings be distributed among the Nightingale Training Schools, and that her 'useful furniture and books' should be placed among the 'matrons, home sisters, ward sisters, nurses and probationers'—an endowment that, in turn, became the basis for the collections at today's Florence Nightingale Museum at St Thomas' Hospital in London.[78] Her concern for personal privacy, however, can be evidenced by her attitude to most of her papers. Her will specifically requested that, with the exception of her India work, 'all my letters, papers and manuscripts … be destroyed without examination'.[79] Why this injunction was never fully carried out is not entirely clear, but there can be no doubt that historians and biographers have been extremely grateful for this oversight, intentional or not.

Similar tensions between personal privacy and public renown accompanied Nightingale when she eventually succumbed on 13 August 1910, in the relatively modest but comfortable room at 10 (formerly 35) South Street in which she had been photographed four years earlier by Elizabeth Bosanquet. Her death was a private affair, attended by just two family members.[80] With her family having declined the government's offer of a ceremony at Westminster Abbey, Nightingale's funeral on Saturday 27 August 1910 in Wellow, Hampshire, was (as *The Times* described) 'of the quietest possible character in accordance with the strongly expressed wish of Miss Nightingale'.[81] In seeking a humble funeral, Nightingale was again a typical representative of her age and class—as Jalland notes, in contrast to the ostentatious funerals of earlier decades, 'by the 1870s simple funerals had become *de rigueur*'.[82] However, Nightingale's funeral did not turn out to be quite as private as intended. Hundreds of members of the public gathered in the rain to witness the approach of her coffin to a ceremony within the Church of St Margaret in East Wellow. British army pallbearers transferred her white-shawled and flower-bedecked coffin from the horse-drawn funeral carriage, and her body's descent into the ground was marked by the Victorian hymn of heroic sacrifice, 'The Son of God Goes Forth to War'.[83] The love of the public and those Nightingale had inspired was marked by a sea of floral tributes and dedications. Her grave avoided any ostentatious eulogy and was simply marked 'F.N' (Fig. 8.3). Yet the vault in which her body was placed—completing a four-sided family tomb that also holds her parents and sister—was a kind of final home, its mortuary architecture resembling a church or house of God. After a life spent among households of faith, Nightingale finally resided in what Barrie calls a 'house tomb', or, in Julian Litten's phrase—perhaps more pertinent given her long periods in the sickroom—an 'eternal bedchamber'.[84]

Nightingale's influence did not, of course, end with her death and burial. Even in the years leading up to it, the reality of her life was dwarfed by the position of her image in the public mind. In 1907, she became the first woman to receive the Order of Merit and, the following year, the second to be granted the Freedom of the City of London, yet these were not so much homages to a living person as acts of commemoration towards someone who had already passed into history. Indeed, Lynn McDonald suggests that had Nightingale's faculties remained fully intact, she would likely have refused such honours.[85] The women who marched in the 1908 procession of the National Union of Women's Suffrage Societies carrying a banner emblazoned 'Florence Nightingale – Crimea' were likewise celebrating an historical figure who just happened to still be alive.[86] These commemorative events instead marked Nightingale's transition into a mythic-like presence in the hearts and minds of the public—a process that was to develop much further after her death.[87]

Fig. 8.3 Photograph of American Red Cross workers placing a wreath on Nightingale's tomb at the Church of St Margaret in East Wellow, Hampshire. Photograph taken c. 1910–1919, received in the American Red Cross photograph collection on 9 September 1919, Library of Congress, Washington, DC, LC-A6195–6721

Notes

1. Thomas Barrie, *House and Home: Cultural Context, Ontological Roles* (London: Routledge, 2017), 55.
2. Deborah Cohen, *Household Gods: The British and Their Possessions* (New Haven: Yale University Press, 2006), 30.
3. See Lynn McDonald, 'Theological Views', in *The Collected Works of Florence Nightingale, Volume 2: Florence Nightingale's Spiritual Journey: Biblical Annotations, Sermons and Journal Notes*, ed. Lynn McDonald (Waterloo, ON: Wilfrid Laurier University Press, 2001) [hereafter *CW* 2], 14–55.
4. McDonald identifies payments to John Smedley's Chapel and Lea Chapel in Derbyshire: see '1877 Diary', *CW* 2, 432.
5. McDonald, 'The Practice of Religion', *CW* 2, 56–90, 88. See also Joann G. Widerquist, 'The Spirituality of Florence Nightingale', *Nursing Research* 41, no. 1 (1992): 49–55.
6. See McDonald, 'The Practice of Religion', *CW* 2, 81.
7. McDonald, 'An Overview of Nightingale's Spiritual Journey', *CW* 2, 5–13.
8. Nightingale to Henry Manning, 30 June 1852, in *The Collected Works of Florence Nightingale, Volume 3: Florence Nightingale's Theology: Essays, Letters and Journal Notes*, ed. Lynn McDonald (Waterloo, ON: Wilfrid Laurier University Press, 2002) [hereafter *CW* 3], 247, 248.
9. Ibid., 248.
10. McDonald, 'Theological Views', *CW* 2, 45.
11. See *The Collected Works of Florence Nightingale, Volume 4: Florence Nightingale on Mysticism and Eastern Religions*, ed. Gérard Vallée (Waterloo, ON: Wilfrid Laurier University Press, 2003) [hereafter *CW* 4].
12. Nightingale to unknown recipient, March 1850, *CW* 4, 440.
13. Nightingale, personal note, undated, *CW* 4, 508.
14. Nightingale, *Suggestions for Thought*, in *The Collected Works of Florence Nightingale, Volume 11: Florence Nightingale's Suggestions for Thought*, ed. Lynn McDonald (Waterloo, ON: Wilfrid Laurier University Press, 2008) [hereafter *CW* 11], 293.
15. Nightingale to Benjamin Jowett, undated, CW 3, 187.
16. Nightingale, Letter 10, 24 November [?1849], *CW* 4, 156.
17. Jill Matus, 'The "Eastern-woman Question": Martineau and Nightingale Visit the Harem', *Nineteenth Century Contexts* 21, no. 1 (1999): 63–87, 65.
18. Nightingale's diary, 17 June 1877, *CW* 2, 456.
19. Nightingale, note of 17 June 1892, *CW* 2, 520.
20. Nightingale, 'The Character of God', unpublished essay [?1871], *CW* 3, 76–96.

21. Rowan Moore, *Why We Build: Power and Desire in Architecture* (New York: HarperCollins, 2013), 65.
22. Barrie, *House and Home*, 55, 67.
23. Nightingale, note of 29 March 1850, *CW* 2, 371.
24. Richard Firth, 'Methodist Worship: With Reference to Historic Practice, The Methodist Worship Book, and Current Patterns in the Newcastle Methodist District', PhD diss., University of Birmingham, 2013, 24, https://etheses.bham.ac.uk/id/eprint/4416/1/Firth13PhD.pdf (accessed 28 April 2020). See also David Hempton, *The Religion of the People: Methodism and Popular Religion, c. 1750–1900* (London: Routledge, 1996).
25. Richard W. Davis, *Dissent in Politics, 1780–1830: The Political Life of William Smith, MP* (London: Epworth, 1971), 148–186.
26. Nightingale's diary, 17 June 1877, *CW* 2, 456. Thanks to Lynn McDonald for email communication on this point.
27. See for example Nightingale's diary, 2 April 1877 and 28 April 1877, *CW* 2, 448, 452.
28. McDonald, 'Bibles Nightingale Gave to Others', *CW* 4, 511–512.
29. Nightingale, notes, 28–29 August 1893 and July 1899, *CW* 2, 531, 552.
30. Nightingale's diary, 17 February 1877, *CW* 2, 441.
31. Nightingale, 'Straight Is the Gate' [draft sermon], undated, *CW* 2, 334.
32. Nightingale, note, undated, 18 August 190[?], *CW* 2, 561.
33. Nightingale, *Suggestions for Thought*, *CW* 11, 291.
34. Ibid., 294.
35. Lynn McDonald, 'Godchildren and Namessakes', in *The Collected Works of Florence Nightingale, Volume 1: Florence Nightingale: An Introduction to Her Life and Family*, ed. Lynn McDonald (Waterloo, ON: Wilfrid Laurier University Press, 2001) [hereafter *CW* 1], 717.
36. See *CW* 1, 717–730.
37. Nightingale to Edmund Verney, 9 June 1881, *CW* 1, 641.
38. Nightingale to Harry Verney, 21 January 1895, *CW* 1, 730; Nightingale to Margaret Verney, 30 September 1896, *CW* 1, 730.
39. Nightingale, annotations to Old Testament extracts, *CW* 2, 164, 168, 148.
40. Nightingale, annotation to Mark 14:13, *CW* 2, 251.
41. Elizabeth Gaskell, letters from October 1854, in *The Letters of Mrs Gaskell*, ed. J. A. V. Chapple and Arthur Pollard (Manchester: Manchester University Press, 1966), 305–322.
42. See Mark Bostridge, *Florence Nightingale: The Making of an Icon* (London: Macmillan, 2008), 393–395.
43. McDonald, '1877 Diary', *CW* 2, 429.
44. Nightingale to her mother, 23 June 1863, *CW* 1, 162.
45. Nightingale to Parthenope, 22 February 1865, *CW* 1, 332.

46. Nightingale to her mother, January 1870, *CW* 1, 194; Nightingale to Harry and Parthenope Verney, 9–10 September 1874, *CW 2*, 12.
47. Nightingale to her mother, January 1870, *CW* 1, 194.
48. Nightingale to Parthenope, 9 July 1886, *CW* 1, 376.
49. Nightingale, draft letter to Jowett [1866], *CW* 2, 354.
50. Jowett to Nightingale, 27 February 1869, quoted in McDonald, 'Sermons', *CW* 2, 325.
51. Jowett to Nightingale, 11 November 1871, quoted in McDonald, 'Sermons', *CW* 2, 326.
52. Nightingale, note to Jowett, undated, *CW 2*, 360.
53. Nightingale, letter [probably to her father], 7 June 1869, *CW* 2, 359.
54. See McDonald, 'Sermons', *CW* 2, 327.
55. Nightingale, 'Lord, to Whom Shall We Go' [draft sermon], *CW* 2, 343–351, 350.
56. Nightingale to Rachel Williams, 23 May 1873, in *The Collected Works of Florence Nightingale, Volume 12: The Nightingale School*, ed. Lynn McDonald (Waterloo, ON: Wilfrid Laurier University Press, 2009) [hereafter *CW* 12], 271.
57. Nightingale, Address 1 [given by Sir Harry Verney] to Nurses of the Nightingale School, London, 8 May 1872, *CW* 12, 763.
58. Nightingale, Address 14 to Nurses of the Nightingale School, 28 May 1900, *CW* 12, 880.
59. Nightingale, Address 2 to Nurses of the Nightingale School, 23 May 1873, *CW* 12, 776.
60. Boyd Hilton, *The Age of Atonement: The Influence of Evangelicalism on Social and Economic Thought, 1785–1865* (Oxford: Oxford University Press, 1993), 335.
61. Pat Jalland, *Death in the Victorian Family* (Oxford: Oxford University Press, 1996), 3. See also Peter C. Jupp and Clare Gittings, eds., *Death in England: An Illustrated History* (Manchester: Manchester University Press, 1999), 237.
62. Barrie, *House and Home*, 3.
63. William Branks, *Heaven Our Home* (1861; Boston: Roberts, 1864), 82–84.
64. See for example Nightingale, diary entry for week of 29 April 1877, *CW* 2, 452.
65. Nightingale's diary, 19 May 1869, *CW* 2, 417.
66. Nightingale to Shore Smith, 16 January 1889, *CW* 3, 210.
67. Nightingale to Buxton, 10 February 1880, *CW* 1, 215.
68. See for example Nightingale to the family of John Cope of Spondon, *Derby Mercury*, 12 April 1855.
69. Nightingale to unnamed recipient, 11 March 1855, *CW* 3, 198.
70. McDonald, 'Introduction', *CW* 2, 32.

71. Jalland, *Death*, 12.
72. Nightingale to her father, 25 March 1853, *CW* 1, 422.
73. Ibid.
74. Nightingale to Sister Frances Wylde, 19 April 1887, *CW* 1, 45.
75. Nightingale's diary, 13–14 December 1877, *CW* 2, 491.
76. Nightingale, Last Will and Codicils, *CW* 1, 852–861, 853.
77. Ibid., 854.
78. Ibid., 857.
79. Ibid.
80. On Victorian deathbed scenes see Jalland, *Death*, 26; Jupp and Gittings, eds., *Death*, 233.
81. *The Times*, 13 August 1910.
82. Jalland, *Death*, 202.
83. McDonald, 'Introduction', *CW* 2, 13. For a detailed account of the funeral procession, burial, and memorials see *The Daily Telegraph*, 22 August 1910, 9–10.
84. Barrie, *House and Home*, 78, ix; Julian Litten, *The English Way of Death: The Common Funeral Since 1450* (Hong Kong: Derek Doyle & Associates, 1991), 195.
85. Lynn McDonald, *Florence Nightingale at First Hand* (London: Bloomsbury, 2010), 26.
86. The banner, designed by Mary Lowndes, is held by the LSE Library and can be viewed at https://vads.ac.uk/large.php?uid=78811 (accessed 28 April 2020).
87. Lara Kriegel, 'On the Death—And Life—Of Florence Nightingale, August 1910', in *BRANCH: Britain, Representation and Nineteenth-Century History,* ed. Dino Franco Felluga, http://www.branchcollective.org/?ps_articles=lara-kriegel-on-the-death-and-life-of-florence-nightingale-august-1910 (accessed 28 April 2020).

CHAPTER 9

Afterlife

There are few contenders to a more enduring cultural afterlife than Florence Nightingale. As the last chapter has shown, the process of her commemoration began long before her death, originating in the abundant images and descriptions that appeared in the new media era of the Crimean War (Chapter 6). This initial 'Nightingale mania' of 1855 and 1856 was to mould and guide subsequent public portrayals in remarkably enduring ways: Nightingale's photographic image on popular visiting cards, her life retold in widely read biographies for girls, and Staffordshire figurines of her displayed on family mantelpieces.[1]

Despite achieving celebrity status within her own lifetime, Nightingale resisted fame, claiming that such publicity was a hindrance to her work and often refusing attempts to capture and reproduce her image.[2] In July 1856, she declined the artist Jerry Barrett's approaches to paint her portrait at Scutari while also acknowledging the futility of attempting to maintain her privacy:

> publicity has been the cause of the greatest drawbacks I have experienced in the prosecution of the work committed to my charge … it is in consequence of this conviction that I have determined in no way to *forward* the making a show of myself or of any person, or thing connected with that work, though I cannot always *prevent* them or me being made a show of.[3]

Even if genuinely felt in 1856, Nightingale had only recently made use of *The Times*' profile to advance, in her words, 'the work committed to

P. Crawford et al., *Florence Nightingale at Home*,
https://doi.org/10.1007/978-3-030-46534-6_9

my charge' (Chapter 6). In the years after the war, she quickly learnt to draw on the power of the newspapers to raise support for her other campaigns. Frustrated at the low profile of the organisation that later became the British Red Cross, in 1870 she recommended that they 'advertise! advertise! advertise!'[4] As such, while the media's incessant focus on her personal image over and above her work made her uncomfortable, Nightingale was no stranger to the advantages of publicity. Her reluctance to appear publicly in the years around the Crimean War was such that journalists, photographers, and artists petitioned her friends and family to satisfy the public demand. 'I hope Mr Sidney Herbert and others of Miss Nightingale's friends will prevail upon her to sit for her portrait', the artist Charles William Knyvett wrote to Nightingale's sister Parthenope in late 1856. While recognising that 'in shrinking from public manifestations she does only what comforts with the dignity of her character', Knyvett pleaded not to be deprived, along with 'thousands of others, who can never see her in the flesh, from the comfort of looking at her likeness'.[5] Nightingale's parents received similar letters and in one case were told that despite

> the unwillingness on the part of Miss Florence Nightingale to allow her portrait to be taken in any way for publication ... thousands of [her] countrymen and countrywomen are desirous of possessing an authentic portrait of her, and I hope she may be induced to alter her determination.[6]

Such were the demands that Nightingale and her family faced during her lifetime. As Lytton Strachey put it in his iconic account of 1918, '[t]he name of Florence Nightingale lives in the memory of the world by virtue of the lurid and heroic adventure of the Crimea'.[7] In writing on the 'cultural afterlife' of the Crimean War, Rachel Bates identifies the growing range of relics and shrines that this Nightingale hagiography brought into being—including an orange purportedly given by Nightingale to a Crimean soldier and displayed today at Claydon House.[8] Photographs, oil paintings, engravings, and drawings of Nightingale graced the walls of homes and galleries.[9] An audio recording of 1890 even captured Nightingale's voice for posterity with a statement that itself contributed to her enduring myth: 'When I am no longer even a memory, just a name, I hope my voice may perpetuate the great work of my life'.[10] After Nightingale's death, her commemoration expanded into the emerging media forms of the early twentieth century. The first film of Nightingale's life appeared in 1915, and many plays, films, and television programmes followed—perhaps most famously the 1951 film *The Lady with a Lamp* starring Anna Neagle. Full-size statues, busts, and memorial plaques appeared in, on, or near buildings and places connected with her life and work, with Arthur George Walker's bronze statue (1915) in Waterloo Place, London, being the best known. Nightingale's image has since featured on coins, banknotes, plates, cups, T-shirts, bath toys, and, from 2020, a barbie doll. A whole genre of Nightingale-themed 'educational manga' has developed in Japan.[11]

In short, Nightingale became a commodity. This is a form of immortality: as Chris Rojek argues, 'once the public face of the celebrity has been elevated and internalized in popular culture, it indeed possesses an immortal quality that permits it to be recycled'.[12] According to Rojek, the posthumous commodification of celebrities is evident in efforts to preserve their homes as tourist attractions that become like shrines.[13] While neither of the Nightingale family homes in Hampshire and Derbyshire are permanent heritage sites, global visitors nonetheless regularly seek permission to visit them. The Florence Nightingale Museum in London (founded in 1989) receives tens of thousands of visitors each year, most from outside the UK. In addition, the Florence Nightingale Foundation, which traces its origins back to the 1855 Nightingale Fund, continues to promote and sponsor nursing education.

Amid this active and wide-ranging cultural afterlife, the question of Nightingale's saintliness has never quite gone away. Terms such as 'angel', 'mystic', 'light', and 'inspiration' characterised her initial representation in October 1854 and have been associated with her image ever since.[14] In 2000, the American Episcopal Church added Nightingale to its list of 'lesser saints', though not without a degree of controversy and objection.[15] Nightingale's importance to the history of nursing combines with her popular association with faith to prompt questions about the ongoing role of religion and spirituality within the profession. Much of the public commemoration of Nightingale now takes place in religious houses, whether at the annual Nightingale memorial service held at Westminster Abbey since 1965, the regular commemorative services in Derbyshire at St Peter's Church and Derby Cathedral, or at similar memorial events at the Church of St Margaret at East Wellow in Hampshire. Yet while Nightingale's career had always intertwined with the British political and social establishment, and, from 1854, the British army, she did not wed herself to these. As noted in the previous chapter, Nightingale deliberately stipulated that Westminster Abbey should not be her place of final rest, and often sought out progressive ways of challenging the norms associated with powerful institutions.

As we completed the manuscript of this book in spring 2020, Britain's relationship both to Nightingale commemoration and to the idea of home took a sudden and unexpected turn. The spread of a new coronavirus around the globe led to billions of people finding themselves in lockdown. People had to adjust rapidly to social isolation and the unsettling sense of being prisoners in their own homes or in temporary accommodation. They had to find creative ways to live their lives; the household and the family took centre stage, rooms were transformed into places of work, exercise, and learning; and the Archbishop of Canterbury even led an Easter Sunday service from his kitchen.[16] Nightingale's life and work, and many of the ideas about home addressed in this book, began to seem like topics to which more people than ever could intimately relate. Not only had Nightingale lived through epidemics, promoted hand washing and healthy home environments, and worked from her bed in seclusion from her family, the celebration of her and her nursing colleagues as military-style heroes anticipated the public applause from members of the public for 'NHS frontline' workers each Thursday evening.[17] The virus forced the cancellation of various planned Nightingale commemoration events in May 2020. London's ExCeL centre, the site of a large-scale nursing conference in Nightingale's honour planned for the autumn, became instead the first of seven NHS Nightingale Hospitals in repurposed spaces, dedicated to providing critical care to victims of the pandemic in England. Its kilometre-long exhibition hall was due to be filled with some 4000 patients laid out in rows, bringing to mind the 'miles of prostrate sick' that Nightingale attended to in the Scutari hospital, itself repurposed to meet the medical emergency of 1854.[18] Although, at the time of writing, these temporary hospitals had admitted far fewer patients than expected—in part due to a lack of nurses trained in critical care—the choice to remember Nightingale in this way demonstrated her continuing presence in the public imagination as a familiar, reassuring figure. The Nightingale name conferred an instant sense of legitimacy and implied that these were not simply improvised field hospitals, but solid institutions in which everything was going to be under control. In the face of the pandemic, the Chief Nursing Officer, Ruth May, spoke of Nightingale as the 'iconic nursing leader of her time', while other nurses echoed the language that she used about the profession by publicly referring to their colleagues as 'work family'.[19] One also suspects that Nightingale might have found these developments to be apposite in her bicentenary year, given that her legacy was to now not only be honoured by a series of academic discussions, but by the prospect of nurses, doctors, and other NHS staff working to save lives in truly exceptional circumstances.

Such ongoing shifts in the patterns of Nightingale commemoration are strikingly appropriate for a figure whose own actions and ideas were neither conventional nor static. Finding Nightingale at home, as this book as has sought to do, has meant following her life and thought to places far less settled than the houses and the institutions into which she was born. Her early life marked a battle to escape the confines and limitations of domesticity, to escape the restrictions forced on her by the conventions of polite society, to act in the world, and to use her intellect for good. However, the same traditions of privilege that frustrated her also offered her a practice of charitable visiting that was to prove formative. Seeing poverty in working-class dwellings drove her lifelong work to bring health and comfort into the mass of homes across Britain and its empire. Yet from an early age, Nightingale realised that she would only ever truly feel at home while engaged in meaningful work. This insight led her to value communities and institutions that could serve as surrogate homes and facilitate women's involvement in the public sphere. She wanted her nurses to belong to a family larger than that contained in a household. Home was not just the bricks and mortar of the places she resided in and visited. For Nightingale, home was a mission: an expansive concept and reality that lay at the heart of her life and thought.

Notes

1. Mark Bostridge, *Florence Nightingale: The Making of an Icon* (London: Macmillan, 2008), xxi, 261; John Plunkett, 'Celebrity and Community: The Poetics of the Carte-de-Visite', *Journal of Victorian Culture* 8, no. 1 (January 2003): 55–79; Martha Vicinus, 'Biographies of Florence Nightingale for Girls', in *Telling Lives in Science: Essays on Scientific Biography*, ed. Michael Shortland and Richard Yeo (Cambridge: Cambridge University Press, 1996), 195–214; Asa Briggs, *Victorian Things* (London: Batsford, 1988), 124–137.
2. Greg Jenner considers Nightingale in his recent popular history of celebrity, *Dead Famous: An Unexpected History of Celebrity from Bronze Age to Silver Screen* (London: Weidenfeld & Nicolson, 2020). For

Nightingale's resistance to her fame, see Bostridge, *Florence Nightingale*, 264–268.

3. Nightingale to Jerry Barrett, 18 July 1856, LMA (FNM) H1/ST/NC3/SU193.
4. Nightingale to Harry Verney, 13 August 1870, in *The Collected Works of Florence Nightingale, Volume 15: Florence Nightingale on Wars and the War Office*, ed. Lynn McDonald (Waterloo, ON: Wilfrid Laurier University Press, 2011), 660.
5. Charles William Kynvett to Parthenope Nightingale, undated [late 1856], Claydon N307.
6. Thomas Hutton to William Nightingale, 19 September 1857, Claydon N307.
7. Lytton Strachey, *Eminent Victorians* (London: Chatto and Windus, 1918), 145.
8. Rachel Bates, 'Curating the Crimea: The Cultural Afterlife of a Conflict', PhD diss., University of Leicester, 2015, 153.
9. For an authoritative list of such images, see Carol Blackett-Ord, 'Florence Nightingale: All Known Portraits', National Portrait Gallery, https://www.npg.org.uk/collections/search/personextended?linkid=mp03298&tab=iconography (accessed 26 February 2020).
10. Florence Nightingale, speech in support of the Light Brigade Relief Fund, 30 July 1890, 2 min., 31 sec.; held by the British Library, 1CD0239287, https://sounds.bl.uk/Accents-and-dialects/Early-spoken-word-recordings/024M-1CD0239287XX-0214V0 (accessed 28 April 2020).
11. Sari Kawana, 'Romancing the Role Model: Florence Nightingale, Shōjo Manga, and the Literature of Self-Improvement', *Japan Review* 23 (2011): 199–223.
12. Chris Rojek, *Celebrity* (London: Reaktion, 2001), 189.
13. Ibid., 59.
14. 'Who Is Mrs Nightingale?', *The Examiner*, 28 October 1854, 682–683.
15. Lynn McDonald, 'The Practice of Religion', in *The Collected Works of Florence Nightingale, Volume 2: Florence Nightingale's Spiritual Journey: Biblical Annotations, Sermons and Journal Notes*, ed. Lynn McDonald (Waterloo, ON: Wilfrid Laurier University Press, 2001), 56–90, 87.
16. Church of England, Easter Sunday Service with the Archbishop of Canterbury, 12 April 2020, https://youtu.be/6bmhRCJ3YAI (accessed 28 April 2020).
17. Richard Bates, 'Florence Nightingale: A Pioneer of Hand Washing and Hygiene for Health', *The Conversation*, 23 March 2020, https://theconversation.com/florence-nightingale-a-pioneer-of-hand-washing-and-hygiene-for-health-134270 (accessed 28 April 2020); Agnes Arnold-Forster and Caitjan Gainty, 'Being Seen as an "NHS Hero" Comes at a Cost', *The*

Times, 21 April 2020, https://www.thetimes.co.uk/article/being-seen-as-an-nhs-hero-comes-at-a-cost-w2q0xzcdx (accessed 30 April 2020).
18. *The Times*, 8 February 1855, 8.
19. Ruth May (Chief Nursing Officer), speech upon the opening of the NHS Nightingale London, 3 April 2020, quoted in 'Nightingale Opens at London's ExCeL Centre', https://www.bbc.co.uk/news/uk-52150598 (accessed 28 April 2020); Amanda Hallums (Chief Nurse, East Kent Hospitals NHS Trust), 3 April 2020, quoted in 'Nurse Aimee O'Rourke Dies After Covid-19 Diagnosis', https://www.bbc.co.uk/news/uk-england-kent-52151231 (accessed 28 April 2020).

Bibliography

Archival Sources

British Library: Nightingale Papers, Canning Papers (BL)
Claydon House Trust: Nightingale Collection, Verney Collection
Derbyshire Record Office (DRO)
Florence Nightingale Museum (FNM)
London Metropolitan Archives (LMA)
Lotherton Hall
University of Nottingham Manuscripts and Special Collections (UoN)
Wellcome Collection

The Collected Works of Florence Nightingale

The set of sixteen volumes is edited by Lynn McDonald, unless otherwise stated, and published by Wilfrid Laurier University Press of Waterloo, ON.

Volume 1: An Introduction to Her Life and Family. 2001.

Volume 2: Florence Nightingale's Spiritual Journey: Biblical Annotations, Sermons and Journal Notes. 2001.

Volume 3: Florence Nightingale's Theology: Essays, Letters and Journal Notes. 2001.

Volume 4: Florence Nightingale on Mysticism and Eastern Religions. Edited by Gérard Vallée. 2003.

Volume 5: Florence Nightingale on Society and Politics, Philosophy, Science, Education and Literature. 2003.

Volume 6: Florence Nightingale on Public Health Care. 2004.

Volume 7: Florence Nightingale's European Travels. 2004.

P. Crawford et al., *Florence Nightingale at Home*,
https://doi.org/10.1007/978-3-030-46534-6

Volume 8: Florence Nightingale on Women, Medicine, Midwifery and Prostitution. 2005.
Volume 9: Florence Nightingale on Health in India. Edited by Gérard Vallée. 2006.
Volume 10: Florence Nightingale on Social Change in India. Edited by Gérard Vallée. 2007.
Volume 11: Florence Nightingale's Suggestions for Thought. 2008.
Volume 12: The Nightingale School. 2009.
Volume 13: Florence Nightingale: Extending Nursing. 2009.
Volume 14: Florence Nightingale: The Crimean War. 2010.
Volume 15: Florence Nightingale on Wars and the War Office. 2011.
Volume 16: Florence Nightingale and Hospital Reform. 2013.

Newspapers and Periodicals

Derby Mercury
Household Words
The Illustrated London News
London Journal
The Morning Chronicle
The Morning Post
Nursing Record
Punch
Quarterly Review
The Daily Telegraph
The Examiner
The Lancet
The London Review
The Philanthropist
The Saturday Review
The Sunday at Home
The Times

Primary Sources

Beecher, Catherine, E. *A Treatise on Domestic Economy: For the Use of Young Ladies at Home and at School*. New York: Harper, 1849.
Beeton, Isabella. *Mrs Beeton's Book of Household Management*. London: S. O. Beeton, 1861.
Benson, Christopher. *The District Visitor's Manual*. London: John W. Parker, 1840.

Branks, William. *Heaven Our Home: We Have No Saviour but Jesus, and No Home but Heaven*. Boston: Roberts, 1864. First published 1861 by William Nimmo.

Bridgeman, Lady Charlotte. *Journals* (1846–57). http://ladycharlottesdiaries.co.uk/Journal/Browse.php. Accessed 25 April 2020.

Burdett-Coutts, Angela. *Woman's Mission: A Series of Congress Papers on the Philanthropic Work of Women*. London: Sampson Low, 1893.

Clare, John. *Poems by John Clare*. Edited by Arthur Symons. London: Frowde, 1908.

Clough, Arthur Hugh. *The Poems and Prose Remains of Arthur Hugh Clough*, vol. 1. Edited by Blanche Athena Clough. London: Macmillan, 1869.

Defoe, Daniel. *A Tour Through the Whole Island of Great Britain*. London: Penguin, 1971. First published in three volumes, 1724–26, by Strahan, Mears, and Stagg.

Downing, Andrew Jackson. *The Architecture of Country Houses*. New York: Da Capo, 1968. First published 1850 by Appleton.

Edge, Frederick Milnes. *A Woman's Example: And a Nation's Work: A Tribute to Florence Nightingale*. London: William Ridgway, 1864.

Eliot, George. *The Mill on the Floss*. Edited by Dinah Birch. Oxford: Oxford University Press, 1996. First published 1860 by William Blackwood.

Fawcett, Millicent Garrett. *Some Eminent Women of Our Times*. London: Macmillan, 1889.

Galton, Douglas. *Observations on the Construction of Healthy Dwellings*. Oxford: Clarendon, 1896.

Gandhi, Mohandas K. 'No. 80, Florence Nightingale'. In *The Collected Works of Mahatma Gandhi*, vol. 5, 1905–6. New Delhi: Ministry of Information and Broadcasting, 1994.

Gaskell, Elizabeth. *The Letters of Mrs Gaskell*. Edited by J. A. V. Chapple and Arthur Pollard. Manchester: Manchester University Press, 1966.

Giffard, Jervis Trigge. *Constance and 'Cap', the Shepherd's Dog: A Reminiscence*. London: Harrison, 1861.

Herbert, Mary Elizabeth. *How I Came Home*. London: Catholic Truth Society, 1894.

Hill, Octavia. *Our Common Land*. London: Macmillan, 1877.

Jennings, L. J., ed. *The Croker Papers*, 2nd ed. London: John Murray, 1885.

Mayhew, Henry. 'Home Is Home, Be It Never So Homely'. In *Meliora: Or, Better Times to Come*, edited by Viscount Ingestre. London: John W. Parker, 1853.

McNeill, John and Tulloch, Alexander. *Report of the Commission of Inquiry into the Supplies of the British Army in the Crimea*. London: Harrison, 1855.

More, Hannah. *Cœlebs in Search of a Wife*. London: Cadell and Davies, 1808.

A New System of Practical Domestic Economy. London: H. Colburn, 1827.

Nightingale, Florence. *Notes on Nursing, and Notes on Nursing for the Labouring Classes: Commemorative Edition with Historical Commentary*. Edited by Victor Skretkowicz. New York: Springer, 2010. First published 1860 and 1861 by Harrison.
Osborne, Sidney Godolphin. *Scutari and Its Hospitals*. London: Dickinson, 1855.
Parkes, Bessie Rayner. 'The Ladies' Sanitary Association', *The English Woman's Journal* 3, no. 14 (1859): 81–82.
Phelps, Elizabeth Stuart. *The Gates Ajar*. Cambridge: Welch and Bigelow, 1873.
Powers, S. R. 'The Diffusion of Sanitary Knowledge'. *Transactions of the National Association for the Promotion of Social Science* (1860): 715–716.
Rathbone, William. *Organization of Nursing: An Account of the Liverpool Nurses' Training School*. Liverpool: Holden, 1865.
Return of Owners of Land: 1873. London: HMSO, 1875.
Richardson, Benjamin Ward. 'Woman as a Sanitary Reformer'. *Transactions of the Sanitary Institute*. London: Sanitary Institute, 1880.
Robertson, Joseph McGregor. *The Household Physician*. London: Blackie, 1890.
Robinson, Frederick. *Diary of the Crimean War*. London: Bentley, 1856.
Sieveking, Edward H. 'On Dispensaries and Allied Institutions'. In *Lectures to Ladies*, edited by F. D Maurice, 91–116. Cambridge: Macmillan, 1855.
Smiles, Samuel. *Self-Help*. London: John Murray, 1859.
———. *Character*. London: John Murray, 1871.

Secondary Sources

Abbott, Mary. *Family Ties: English Families 1540–1920*. London: Routledge, 1993.
Ablow, Rachel. 'Harriet Martineau and the Impersonality of Pain'. *Victorian Studies* 56, no. 4 (2014): 675–697.
Aidt, Toke S. and Franck, Raphaël. 'Democratisation Under the Threat of Revolution: Evidence from the Great Reform Act of 1832'. *Econometrica* 83, no. 2 (2015): 505–547.
Allen, Michelle. 'From Cesspool to Sewer: Sanitary Reform and the Rhetoric of Resistance, 1848–1880'. *Victorian Literature and Culture* 30, no. 2 (2002): 383–402.
Anderson, Olive. *A Liberal State at War*. London: Macmillan, 1967.
Anstruther, Ian. *The Scandal of the Andover Workhouse*. London: Bles, 1973.
Arnold, David. *Colonizing the Body: State Medicine and Epidemic Disease in Nineteenth Century India*. Berkeley: University of California Press, 1993.
Bachelard, Gaston. *The Poetics of Space*. Translated by Maria Jolas. Boston: Beacon, 1994.

Bailin, Miriam. *The Sickroom in Victorian Fiction: The Art of Being Ill*. Cambridge: Cambridge University Press, 1994.

Ballantyne, Tony and Burton, Antoinette. *Empires and the Reach of the Global, 1870–1945*. Cambridge, MA: Harvard University Press, 2012.

Baly, Monica. *Florence Nightingale and the Nursing Legacy*. Beckenham: Croom Helm, 1986.

Barker, Hannah. *Family and Business During the Industrial Revolution*. Oxford: Oxford University Press, 2017.

Barrell, John. 'Death on the Nile: Fantasy and the Literature of Tourism 1840–1860'. *Essays in Criticism* 41, no. 2 (1991): 97–127.

Barrie, Thomas. *House and Home: Cultural Contexts, Ontological Roles*. London: Routledge, 2017.

Bashford, Alison. *Purity and Pollution*. London: Macmillan, 1998.

———. *Imperial Hygiene: A Critical History of Colonialism, Nationalism and Public Health*. Houndsmills: Palgrave, 2004.

Bates, Crispin and Carter, Marina, eds. *Mutiny at the Margins: New Perspectives on the Indian Uprising of 1857: Volumes I–VII: Documents of the Indian Uprising*. New Delhi: Sage, 2013–2017.

Bates, Rachel. 'Curating the Crimea: The Cultural Afterlife of a Conflict'. PhD diss., University of Leicester, 2015.

———. '"All Touched My Hand": Queenly Sentiment and Royal Prerogative'. *19: Interdisciplinary Studies in the Long Nineteenth Century* 20 (13 May 2015).

Bates, Richard. 'Florence Nightingale: A Pioneer of Hand Washing and Hygiene for Health'. *The Conversation*, 23 March 2020. https://theconversation.com/florence-nightingale-a-pioneer-of-hand-washing-and-hygiene-for-health-134270. Accessed 28 April 2020.

Bektas, Yakup. 'The Crimean War as a Technological Enterprise'. *Notes and Records: The Royal Society Journal of the History of Science* 71, no. 3 (2017): 233–262.

Bell, Duncan. *The Idea of Greater Britain: Empire and the Future of World Order, 1860–1900*. Princeton: Princeton University Press, 2011.

Black, Barbara J. *A Room of His Own: A Literary-Cultural Study of Victorian Clubland*. Athens, OH: Ohio University Press, 2012.

Black, Jeremy. *Italy and the Grand Tour*. New Haven: Yale University Press, 2003.

Blackett-Ord, Carol. 'Florence Nightingale: All Known Portraits'. National Portrait Gallery, 26 February 2020. https://www.npg.org.uk/collections/search/personextended?linkid=mp03298&tab=iconography. Accessed 29 April 2020.

Blunt, Alison. 'Cultural Geography: Cultural Geographies of Home'. *Progress in Human Geography* 9, no. 4 (2005): 505–515.

Blunt, Alison and Dowling, Robyn M. *Home*. London: Routledge, 2006.
Blunt, Alison and Varley, Ann. 'Geographies of Home'. *Cultural Geographies* 11 (2004): 3–6.
Boardman, Kay. 'The Ideology of Domesticity: The Regulation of the Household Economy in Victorian Women's Magazines'. *Victorian Periodicals Review* 33, no. 2 (2000): 150–164.
Booth, Alison. 'A Bestiary of Florence Nightingales: Strachey and Collective Biographies of Women'. *Victorian Studies* 61, no. 1 (2018): 93–98.
Bostridge, Mark. *Florence Nightingale: The Woman and Her Legend*. London: Penguin, 2009.
———. 'Self-Sacrifice and Parthenope Nightingale'. *Times Literary Supplement*, 24 May 2002.
Bourdieu, Pierre. *The Logic of Practice*. Translated by Richard Nice. Cambridge: Polity, 1990.
———. *The Social Structures of the Economy*. Translated by Chris Turner. Cambridge: Polity, 2005.
Branca, Patricia. *Silent Sisterhood: Middle-Class Women in the Victorian Home*. London: Croom Helm, 1975.
Braudy, Leo. *The Frenzy of Renown: Fame and Its History*. Oxford: Oxford University Press, 1986.
Briggs, Asa. *Victorian Things*. London: Batsford, 1988.
Burton, Elizabeth. *The Early Victorians at Home, 1837–1861*. London: Longman, 1972.
Buxton, Doreen and Charlton, Christopher. *Cromford Revisited*, Nottingham: Derwent Valley Mills World Heritage Site Educational Trust, 2013.
Calabria, Michael D. and Macrae, Janet A., eds. *Suggestions for Thought by Florence Nightingale: Selections and Commentaries*. Philadelphia: University of Pennsylvania Press, 1994.
Cartwright, F. F. 'Miss Nightingale's Dearest Friend'. *Proceedings of the Royal Society of Medicine* 69, no. 3 (March 1976): 169–175.
Cashmore, Ellis. *Celebrity/Culture*. Abingdon: Routledge, 2006.
Cavaliero, Roderick. *Italia Romantica: English Romantics and Italian Freedom*. London: I.B. Tauris, 2005.
Chapman, Stanley, ed. *The History of Working-Class Housing*. Newton Abbot: David and Charles, 1971.
———. 'Peter Nightingale, Richard Arkwright, and the Derwent Valley Cotton Mills, 1771–1818'. *Derbyshire Archaeological Journal* 133 (2013): 166–188.
Cohen, Deborah. *Household Gods: The British and Their Possessions*. New Haven: Yale University Press, 2006.
Cohen, Monica Feinberg. *Professional Domesticity in the Victorian Novel: Women, Work and Home*. Cambridge: Cambridge University Press, 1998.
Conacher, J. B. *Britain and the Crimea, 1855–56*. Basingstoke: Macmillan, 1987.

Cook, Edward. *The Life of Florence Nightingale*, vols. 1 and 2. London: Macmillan, 1913.
Cooper, Fredrick and Stoler, Ann Laura, eds. *Tensions of Empire: Colonial Cultures in a Bourgeois World*. Berkeley: University of California Press, 1997.
Corbett, Mary Jean. 'Cousin Marriage, Then and Now'. *Victorian Review* 39, no. 2 (2013): 74–78.
Crook, Tom. *Governing Systems: Modernity and the Making of Public Health in England, 1830–1910*. Oakland: University of California Press, 2016.
Cuming, Emily. *Housing, Class and Gender in Modern British Writing, 1880–2012*. Cambridge: Cambridge University Press, 2016.
Darwin, John. *The Empire Project: The Rise and Fall of the British World-System 1830–1970*. Cambridge: Cambridge University Press, 2009.
Daunton, Martin James. *House and Home in the Victorian City*. London: Edward Arnold, 1983.
Davidoff, Leonore and Hall, Catherine. *Family Fortunes: Men and Women of the English Middle Class 1780–1850*, 3rd ed. London: Routledge, 2019.
Davies, Celia. 'The Health Visitor as Mother's Friend: A Woman's Place in Public Health, 1900–14'. *Social History of Medicine* 1, no. 1 (1988): 39–59.
Davis, Richard W. *Dissent in Politics 1780–1830: The Political Life of William Smith, MP*. London: Epworth, 1971.
Delamont, Sara. 'The Contradictions in Ladies' Education'. In *The Nineteenth-Century Woman: Her Cultural and Physical World*, edited by Lorna Duffin and Sara Delamont, 134–187. London: Croom Helm, 1978.
Dossey, Barbara M. 'Florence Nightingale: Her Crimean Fever and Chronic Illness'. *Journal of Holistic Nursing* 28, no. 1 (2010): 38–53.
Dutton, Ralph. *The Victorian Home*. Twickenham: Bracken, 1954.
Elliott, Dorice Williams. *The Angel Out of the House: Philanthropy and Gender in Nineteenth-Century England*. Charlottesville: University of Virginia Press, 2002.
Evans, Eric J. *The Forging of the Modern State, Early Industrial Britain, 1873–c.1870*, 4th ed. Oxford: Routledge, 2019.
Figes, Orlando. *Crimea: The Last Crusade*. London: Allen Lane, 2011.
Firth, Richard. 'Methodist Worship: With Reference to Historic Practice, the Methodist Worship Book, and Current Patterns in the Newcastle Methodist District'. PhD diss., University of Birmingham, 2013.
Fitton, R. S. *The Arkwrights: Spinners of Fortune*. Manchester: Manchester University Press, 1989.
Flanders, Judith. *The Victorian House: Domestic Life from Childbirth to Deathbed*. London: Harper, 2003.
———. *The Making of Home*. London: Atlantic, 2014.
Flint, Kate. *The Victorians and the Visual Imagination*. Cambridge: Cambridge University Press, 2000.

Foucault, Michel. *Discipline and Punish: The Birth of the Prison*. Translated by Michael Sheridan. New York: Pantheon, 1977.

Fox, Michael Allen. *Home: A Very Short Introduction*. Oxford: Oxford University Press, 2016.

Freedgood, Elaine. *Victorian Writing about Risk: Imagining a Safe England in a Dangerous World*. Cambridge: Cambridge University Press, 2009.

Furneaux, Holly. *Military Men of Feeling: Emotion, Touch, and Masculinity in the Crimean War*. Oxford: Oxford University Press, 2016.

Gagnier, Regenia. *Subjectivities: A History of Self-Representation in Britain, 1832–1920*. Oxford: Oxford University Press, 1991.

Gamarnikow, Eva. 'Nurse or Woman: Gender and Professionalism in Reformed Nursing, 1860–1923'. In *Anthropology and Nursing*, edited by Pat Holden and Jenny Littleworth, 110–129. London: Routledge, 1991.

Gan, Wendy. 'Solitude and Community: Virginia Woolf, Spatial Privacy and A Room of One's Own'. *Literature & History* 18, no. 1 (2009): 68–80.

Garvin, James Louis. *The Life of Joseph Chamberlain*, vol. 2. London: Macmillan, 1933–1935.

Gauldie, Enid. *Cruel Habitations: A History of Working-Class Housing 1780–1918*. London: Allen and Unwin, 1974.

George, Rosemary Marangoly. *The Politics of Home: Postcolonial Relocations and Twentieth Century Fiction*. Cambridge: Cambridge University Press, 1996.

Gerard, Jessica. *Country House Life: Families and Servants 1815–1914*. Oxford: Blackwell, 1994.

———. 'Lady Bountiful: Women of the Landed Classes and Rural Philanthropy'. *Victorian Studies* 30, no. 2 (1987), 183–210.

Gill, Gillian. *Nightingales: Florence and Her Family*. London: Hodder and Stoughton, 2004.

Girouard, Mark. *Life in the English Country House: A Social and Architectural History*. New Haven: Yale University Press, 1978.

Gleadle, Kathryn. *The Early Feminists: Radical Unitarians and the Emergence of the Women's Rights Movement, 1831–51*. New York: St Martin's Press, 1995.

Gloag, John. *Victorian Comfort: A Social History of Design from 1830–1900*. Newton Abbott: David and Charles, 1973.

Godden, Judith. 'The Dream of Nursing the Empire'. In *Notes on Nightingale: The Influence and Legacy of a Nursing Icon*, edited by Sioban Nelson and Anne Marie Rafferty, 55–75. Ithaca: Cornell University Press, 2010.

Goldstein, Jan. 'Professional Knowledge and Professional Self-Interest: The Rise and Fall of Monomania in 19th-Century France'. *International Journal of Law and Psychiatry* 21, no. 4 (1998): 385–396.

Goodman, Ruth. *How to Be a Victorian*. London: Penguin, 2014.

Gorman, Daniel. *Imperial Citizenship: Empire and the Question of Belonging*. Manchester: Manchester University Press, 2007.

Gorsky, Martin. *Patterns of Philanthropy: Charity and Society in Nineteenth-Century Bristol*. Woodbridge: Boydell Press, 1999.
Gottlieb, Beatrice. *The Family in the Western World: From the Black Death to the Industrial Age*. Oxford: Oxford University Press, 1993.
Gourlay, Jharna. *Florence Nightingale and the Health of the Raj*. Aldershot: Ashgate, 2003.
Griffin, Carl J. 'The Violent Captain Swing?' *Past & Present* 209, no. 1 (2010): 149–180.
Groth, Helen. 'Technological Mediations and the Public Sphere: Roger Fenton's Crimea Exhibition and the "Charge of the Light Brigade"'. *Victorian Literature and Culture* 30, no. 2 (September 2002): 553–570.
Guha, Ranjit. 'Not at Home in Empire'. *Critical Inquiry* 23, no. 3 (1997): 482–493.
Gust, Onni. '"The Perilous Territory of Not Belonging": Exile and Empire in Sir James Mackintosh's Letters from Early Nineteenth-Century Bombay'. *History Workshop Journal* 86 (2018): 22–43.
Haldane, Elizabeth Sanderson. *Mrs. Gaskell and Her Friends*. London: Hodder and Stoughton, 1930.
Hamlett, Jane. *At Home in the Institution: Material Life in Asylums, Lodging Houses and Schools in Victorian and Edwardian England*. Basingstoke: Palgrave, 2015.
———. *Material Relations: Domestic Interiors and Middle-Class Families in England, 1850–1910*. Manchester: Manchester University Press, 2010.
Hamlin, Christopher. *Public Health and Social Justice in the Age of Chadwick: Britain, 1800–1854*. Cambridge: Cambridge University Press, 1998.
Hardy, Gwen. *William Rathbone and the Early History of District Nursing*. Ormskirk: Hesketh, 1981.
Harling, Robert. *Home: A Victorian Vignette*. London: Constable, 1938.
Harrison, Mark. *Public Health in British India: Anglo-Indian Preventative Medicine, 1859–1914*. Cambridge: Cambridge University Press, 1994.
Harvey, Karen. *The Little Republic: Masculinity and Domestic Authority in Eighteenth-Century Britain*. Oxford: Oxford University Press, 2012.
Hawkins, Sue. *Nursing and Women's Labour in the Nineteenth Century: The Quest for Independence*. Abingdon: Routledge, 2010.
Heathcote, Edwin. *The Meaning of Home*. London: Frances Lincoln, 2014.
Helmstadter, Carol and Godden, Judith. *Nursing Before Nightingale, 1815–1899*. Farnham: Ashgate, 2011.
Hewitt, Martin. 'District Visiting and the Constitution of Domestic Space in the Mid-Nineteenth Century'. In *Domestic Space: Reading the Nineteenth-Century Interior*, edited by Ingra Bryden and Janet Floyd, 121–141. Manchester: Manchester University Press, 1999.
Hey, David. *Derbyshire: A History*. Lancaster: Carnegie, 2008.

Hibbert, Christopher. *The Great Mutiny: India 1857*. Harmondsworth: Penguin, 1978.

———. *The Destruction of Lord Raglan*. London: Longmans, 1961.

Hilton, Boyd. *The Age of Atonement: The Influence of Evangelicalism on Social and Economic Thought, 1785–1865*. Oxford: Oxford University Press, 1993.

Hobbs, Andrew. *A Fleet Street in Every Town: The Provincial Press in England, 1855–1900*. Cambridge: Open Book, 2018.

Horn, Pamela. *Ladies of the Manor: Wives and Daughters in Country-House Society 1830–1918*. Stroud: Allan Sutton, 1997.

Houston, Natalie M. 'Reading the Victorian Souvenir: Sonnets and Photographs of the Crimean War'. *Yale Journal of Criticism* 14, no. 2 (2001): 353–383.

Howse, Carrie. 'From Lady Bountiful to Lady Administrator: Women and the Administration of Rural District Nursing in England, 1880–1925'. *Women's History Review* 15, no. 3 (2006): 423–441.

Hunter, J. Paul. *Before Novels: The Cultural Contexts of Eighteenth-Century English Fiction*. New York: Norton, 1990.

Huntsman, Richard G., Bruin, Mary, and Holttum, Deborah. 'Twixt Candle and Lamp: The Contribution of Elizabeth Fry and the Institution of Nursing Sisters to Nursing Reform'. *Medical History* 46 (2002): 351–380.

Ichikawa, Cieko. 'Writing as a Female National and Imperial Responsibility: Florence Nightingale's Scheme for Social and Cultural Reforms in England and India'. *Victorian Literature and Culture* 39 (2011): 87–105.

Jalland, Pat. *Death in the Victorian Family*. Oxford: Oxford University Press, 1996.

Jordan, Ellen. *The Women's Movement and Women's Employment in Nineteenth Century Britain*. London: Routledge, 1999.

Joyce, Patrick. *The State of Freedom: A Social History of the British State Since 1800*. Cambridge: Cambridge University Press, 2013.

Jupp, Peter C. and Gittings, Clare, eds. *Death in England: An Illustrated History*. Manchester: Manchester University Press, 1999.

Kaminsky, Uwe. 'German "Home Mission" Abroad: The *Orientarbeit* of the Deaconess Institution Kaiserswerth in the Ottoman Empire'. In *New Faith in Ancient Lands: Western Mission in the Middle East in the Nineteenth and Early Twentieth Centuries*, edited by Heleen Murre-van den Berg, 191–209. Leiden: Brill, 2006.

Keen, Norman. *Florence Nightingale*. Derby: J. H. Hall, 1982.

Keller, Ulrich. *The Ultimate Spectacle*. Amsterdam: Gordon and Breach, 2001.

Kertzer, David I. and Barbagli, Marzio. *Family Life in the Long Nineteenth Century, 1789–1913*. New Haven: Yale University Press, 2002.

Kinglake, A. W. *The Invasion of the Crimea*, vol. 6. London: Blackwood, 1885.

Kriegel, Lara. 'On the Death—And Life—Of Florence Nightingale, August 1910'. In *BRANCH: Britain, Representation and Nineteenth-Century*

History, edited by Dino Franco Felluga. http://www.branchcollective.org/?ps_articles=lara-kriegel-on-the-death-and-life-of-florence-nightingale-august-1910. Accessed 25 February 2020.
Kuper, Adam. *Incest and Influence: The Private Life of Bourgeois England*. Cambridge, MA: Harvard University Press, 2009.
Lambert, Andrew and Badsey, Stephen. *The War Correspondents: The Crimean War*. Stroud: Sutton, 1994.
Laver, James. *The Liberty Story*. London: Liberty, 1959.
Levitan, Kathrin. 'Redundancy, the "Surplus Woman" Problem, and the British Census, 1851–1861'. *Women's History Review* 17, no. 3 (2008): 359–376.
Lewis, Richard Albert. *Edwin Chadwick and the Public Health Movement 1832–1854*. London: Longman, 1952.
Litten, Julian. *The English Way of Death: The Common Funeral Since 1450*. Hong Kong: Derek Doyle, 1991.
Lochhead, Marion. *The Victorian Household*. London: John Murray, 1964.
Logan, Deborah Anna, ed. *Lives of Victorian Political Figures III, Volume 2: Florence Nightingale*. London: Pickering and Chatto, 2008.
Logan, Thad. *The Victorian Parlour*. Cambridge: Cambridge University Press, 2001.
Lucas, John. *John Clare*. Plymouth: Northcote, 1994.
McClintock, Anne. *Imperial Leather: Race, Gender and Sexuality in the Colonial Context*. London: Routledge, 1995.
McDonald, Lynn. *Florence Nightingale at First Hand*. London: Bloomsbury, 2010.
———. *Florence Nightingale: Nursing and Healthcare Today*. New York: Springer, 2018.
McKeon, Michael. *The Secret History of Domesticity: Public, Private and the Division of Knowledge*. Baltimore: Johns Hopkins University Press, 2007.
MacKinnon, Aran S. and Ablard, Jonathan D., eds. *(Un)Healthy Interiors: Contestations at the Intersection of Public Health and Private Space*. Carrollton, GA: University of West Georgia, 2005.
Mackintosh, J. M. *Housing and Family Life*. London: Cassell, 1952.
MacQueen, Joyce Schroeder. 'Florence Nightingale's Nursing Practice'. *Nursing History Review* 15 (2007): 29–49.
McWilliam, Rohan. 'The Theatricality of the Staffordshire Figurine'. *Journal of Victorian Culture* 10, no. 1 (2005): 107–114.
Magnello, Eileen. 'Victorian Statistical Graphics and the Iconography of Florence Nightingale's Polar Area Graph'. *BSHM Bulletin: Journal of the British Society for the History of Mathematics* 27, no. 1 (2012): 13–37.
Maguire, Hugh. 'The Victorian Theatre as a Home from Home'. *Journal of Design History* 13, no. 2 (2000): 107–121.

Marcus, Sharon. *Apartment Stories: City and Home in Nineteenth-Century Paris and London*. Berkeley: University of California Press, 1999.

Markovits, Stefanie. *The Crimean War in the British Imagination*. Cambridge: Cambridge University Press, 2009.

Mathieson, Charlotte. *Mobility in the Victorian Novel: Placing the Nation*. Basingstoke: Palgrave, 2015.

Matus, Jill. 'The "Eastern-woman Question": Martineau and Nightingale Visit the Harem'. *Nineteenth Century Contexts* 21, no. 1 (1999): 63–87.

Maynes, Mary Jo and Waltner, Ann. *The Family: A World History*. Oxford: Oxford University Press, 2012.

Metz, Nancy Aycock. 'Discovering a World of Suffering: Fiction and the Rhetoric of Sanitary Reform'. *Nineteenth-Century Contexts* 15, no. 1 (1991): 65–81.

Morley, David. *Home Territories: Media, Mobility and Identity*. London: Routledge, 2000.

Moore, Judith. *A Zeal for Responsibility: The Struggle for Professional Nursing in Victorian England, 1868–1883*. Athens: University of Georgia Press, 1988.

Moore, Rowan. *Why We Build: Power and Desire in Architecture*. New York: Harper Design, 2013.

Nagai, Kaori. 'Florence Nightingale and the Irish Uncanny'. *Feminist Review* 77 (2004): 26–45.

Nelson, Claudia. *Family Ties in Victorian England*. London: Praeger, 2007.

Nicholson, Shirley. *A Victorian Household*. Thrupp: Sutton, 1998.

Nixon, Kirsteen. *The World of Florence Nightingale*. London: Pitkin, 2011.

Nolte, Karen. 'Protestant Nursing Care in Germany in the 19th Century: Concepts and Social Practice'. In *Routledge Handbook on the Global History of Nursing*, edited by Patricia D'Antonio, Julie A. Fairman, and Jean C. Whelan, 167–182. Abingdon: Routledge, 2013.

Oliver, Hermia. 'The Shore Smith Family Library: Arthur Hugh Clough and Florence Nightingale'. *Book Collector* 28 (1979): 521–529.

O'Malley, Ida. *Florence Nightingale 1820–1856: A Study of Her Life Down to the End of the Crimean War*. London: Butterworth, 1931.

Orme, Edward B. 'A History of the Illustrated London News' (1986). https://www.iln.org.uk/iln_years/historyofiln.htm. Accessed 29 April 2020.

Partridge, Amy. 'Public Health for the People, the Use of Exhibition and Performance to Stage the "Sanitary Idea" in Victorian Britain'. PhD diss., Northwestern University, 2005.

Peck, John. *War, the Army and Victorian Literature*. Basingstoke: Palgrave, 1998.

Penner, Louise. *Victorian Medicine and Social Reform: Florence Nightingale Among the Novelists*. New York: Palgrave, 2010.

Peterson, Jeanne. *Family, Love, and Work in the Lives of Victorian Gentlewomen*. Bloomington: Indiana University Press, 1989.

Pickering, George. *Creative Malady: Illness in the Lives and Minds of Charles Darwin, Florence Nightingale, Mary Baker Eddy, Sigmund Freud, Marcel Proust, Elizabeth Barrett Browning*. Oxford: Oxford University Press, 1974.
Ponsonby, Margaret. *Studies from Home: English Domestic Interiors, 1750–1850*. Aldershot: Ashgate, 2007.
Plotz, John. *Portable Property: Victorian Culture on the Move*. Princeton: Princeton University Press, 2008.
Plunkett, John. 'Celebrity and Community: The Poetics of the *Carte-de-Visite*'. *Journal of Victorian Culture* 8, no. 1 (January 2003): 55–79.
Poovey, Mary. *Uneven Developments: The Ideological Work of Gender in Mid-Victorian England*. Chicago: University of Chicago Press, 1988.
———. *Making a Social Body: British Cultural Formation, 1830–1864*. Chicago: University of Chicago Press, 1995.
Prochaska, Frank. 'Women in English Philanthropy, 1790–1830'. *International Review of Social History* 19, no. 3 (1974): 426–445.
———. *Women and Philanthropy in Nineteenth-Century England*. Oxford: Clarendon Press, 1980.
Pugh, Evelyn L. 'Florence Nightingale and J. S. Mill Debate Women's Rights'. *Journal of British Studies* 21, no. 2 (Spring 1982): 118–138.
Pryor, Francis. *Home: A Time Traveller's Tales from Britain's Prehistory*. London: Penguin, 2015.
Ramanna, Mridula. 'Florence Nightingale and the Bombay Presidency'. *Social Scientist* 30, no. 9 (2002): 31–46.
Reeves, Caroline. 'The Modern Woman as Global Exemplar, Part II: Florence Nightingale and a Transnational Gendered Modernity'. *History Compass* 13, no. 4 (2015), 191–200.
Relph, Edward. *Place and Placelessness*. London: Pion, 1976.
Rendall, Jane. 'The Condition of Women, Women's Writing and the Empire in Nineteenth Century Britain'. In *At Home with Empire: Metropolitan Culture and the Imperial World*, edited by Catherine Hall and Sonya Rose, 101–121. Cambridge: Cambridge University Press, 2006.
Reynolds, K. D. *Aristocratic Women and Political Society in Victorian Britain*. Oxford: Clarendon, 1998.
Ridley, Sarah. *Florence Nightingale: Social Reformer and Pioneer of Nursing*. London: Franklin Watts, 2020.
Rogaski, Ruth. *Hygienic Modernity: Meanings of Health and Disease in Treaty-Port China*. Berkeley: University of California Press, 2004.
Rojek, Chris. *Celebrity*. London: Reaktion, 2001.
Rosenberg, Charles E. *Explaining Epidemics and Other Studies in the History of Medicine*. Cambridge: Cambridge University Press, 1992.
Rubenhold, Hallie. *The Five: The Untold Lives of the Women Killed by Jack the Ripper*. London: Penguin, 2019.

Rubenstein, David. *Victorian Homes*. Newton Abbott: David & Charles, 1974.
Ruiz, Marie. *British Female Emigration Societies and the New World, 1860–1914*. London: Palgrave, 2017.
Rybczynski, Witold. *Home: A Short History of an Idea*. New York: Penguin, 1987.
Sanders, Andrew, ed. *Great Victorian Lives: An Era in Obituaries*. London: Times Books, 2007.
Sattin, Anthony. *A Winter on the Nile: Florence Nightingale, Gustave Flaubert and the Temptations of Egypt*. London: Hutchinson, 2010.
Schaffer, Talia. *Novel Craft: Victorian Domestic Handicraft and Nineteenth-Century Fiction*. Oxford: Oxford University Press, 2011.
Searle, Geoffrey Russell. *Entrepreneurial Politics in Mid-Victorian Britain*. Oxford: Oxford University Press, 1993.
Selanders, Louise C. and Crane, Patrick. 'Florence Nightingale in Absentia: Nursing and the 1893 Columbian Exposition'. *Journal of Holistic Nursing* 28, no. 4 (2010): 305–312.
Seymer, Lucy. *Florence Nightingale's Nurses: The Nightingale Training School*. London: Pitman, 1960.
Showalter, Elaine. 'Florence Nightingale's Feminist Complaint: Women, Religion and "Suggestions for Thought"'. *Signs* 6, no. 3 (1981): 395–412.
———. *The Female Malady: Women, Madness, and English Culture, 1830–1980*. London: Virago, 1987.
Shuttleworth, Sally. *Charlotte Brontë and Victorian Psychology*. Cambridge: Cambridge University Press, 1996.
Sinha, Mrinalini. *Colonial Masculinity: The 'Manly Englishman' and the 'Effeminate Bengali' in the Later Nineteenth Century*. Manchester: Manchester University Press, 1995.
Smith, Frances Barrymore. *Florence Nightingale: Reputation and Power*. London: Croom Helm, 1982.
Snyder, Katherine V. 'From Novel to Essay: Gender and Revision in Florence Nightingale's "Cassandra"'. In *The Politics of the Essay: Feminist Perspectives*, edited by Ruth-Ellen B. Joeres and Elizabeth Mittman, 23–40. Bloomington: Indiana University Press, 1993.
Spacks, Patricia Meyer. 'The Privacy of the Novel'. *NOVEL: A Forum on Fiction* 31, no. 3 (July 1998): 304–316.
Spivak, Gayatri. 'Three Women's Texts and a Critique of Imperialism'. *Critical Inquiry* 12, no. 1 (1985): 243–261.
Stanley, Liz. 'The Epistolarium: On Theorizing Letters and Correspondences'. *Auto/Biography* 12 (2004): 201–235.
Stanmore, Arthur Hamilton-Gordon. *Sidney Herbert, Lord Herbert of Lea*, vol. 1. London: John Murray, 1906.

Steedman, Carolyn. 'What a Rag Rug Means'. In *Domestic Space: Reading the Nineteenth-Century Interior*, edited by Ingra Bryden and Janet Floyd, 18–39. Manchester: Manchester University Press, 1999.

Stocks, Mary D. *A Hundred Years of District Nursing*. London: Allen & Unwin, 1960.

Stotsky, Sandra. 'Writing in a Political Context: The Value of Letters to Legislators'. *Written Communication* 4 (1987): 394–410.

Strachey, Lytton. *Eminent Victorians*. London: Chatto and Windus, 1918.

Strachey, Ray. *'The Cause': A Short History of the Women's Movement in Great Britain*. London: G. Bell, 1928.

Summers, Anne. 'A Home from Home: Women's Philanthropic Work in the Nineteenth Century'. In *Fit Work for Women*, edited by Sandra Burman, 33–63. London: Croom Helm, 1979.

———. 'The Mysterious Demise of Sarah Gamp: The Domiciliary Nurse and Her Detractors, c. 1830–1860'. *Victorian Studies* 32, no. 3 (1989): 365–86.

Sweet, Rosemary. *Cities and the Grand Tour: The British in Italy, c. 1690–1820*. Cambridge: Cambridge University Press, 2012.

Tate, Trudy. 'On Not Knowing Why: Memorializing the Light Brigade'. In *Literature, Science, Psychoanalysis, 1830–1970: Essays in Honour of Gillian Beer*, edited by Helen Small and Trudi Tate, 160–180. Oxford: Oxford University Press, 2003.

Teukolsky, Rachel. 'Novels, Newspapers, and Global War: New Realisms in the 1850s'. *Novel* 45, no. 1 (May 2012): 31–55.

Tosh, John. *A Man's Place: Masculinity and the Middle-Class Home in Victorian England*. New Haven: Yale University Press, 1999.

Veysey, Iris. 'A Statistical Campaign: Florence Nightingale and Harriet Martineau's *England and Her Soldiers*'. *Science Museum Group Journal* 5 (2016).

Vicinus, Martha. *Independent Women: Work and Community for Single Women, 1850–1920*. Chicago: University of Chicago Press, 1985.

———. 'Biographies of Florence Nightingale for Girls'. In *Telling Lives in Science: Essays on Scientific Biography*, edited by Michael Shortland and Richard Yeo, 195–214. Cambridge: Cambridge University Press, 1996.

Wake, Roy. *The Nightingale Training School 1860–1996*. London: Haggerston, 1998.

Wald, Priscilla. *Contagious: Cultures, Carriers, and the Outbreak Narrative*. Durham, NC: Duke University Press, 2008.

Wallace, Anne D. *Sisters and the English Household: Domesticity and Women's Autonomy in Nineteenth-Century English Literature*. New York: Anthem Press, 2018.

Widerquist, Joann, G. 'The Spirituality of Florence Nightingale'. *Nursing Research* 41, no. 1 (1992): 49–55.

Wigglesworth, George. *Florence Nightingale's Journey Home from the Crimea*. 2010. http://www.wigglesworth.me.uk/local_history/pdf/Florence's%20journey%20home.PDF. Accessed 29 April 2020.

Wohl, Anthony. *Endangered Lives*. London: Dent, 1983.

Wood, Andy. *The Politics of Social Conflict: The Peak Country, 1520–1770*. Cambridge: Cambridge University Press, 1999.

Woodham-Smith, Cecil. *Florence Nightingale 1820–1910*. London: Constable, 1950.

Worsley, Lucy. *Queen Victoria: Daughter, Wife, Mother, Widow*. London: Hodder and Stoughton, 2018.

Yeo, Eileen Janes. 'Social Motherhood and the Sexual Communion of Labour in British Social Science, 1850–1950'. *Women's History Review* 1, no. 1 (1992): 63–87.

Yorke, Trevor. *The Victorian House Explained*. Newbury: Countryside, 2005.

Young, D. A. B. 'Florence Nightingale's Fever'. *British Medical Journal* 311 (December 1995): 1697–1700.

Young, Iris Marion. 'House and Home: Feminist Variations on a Theme'. In *Intersecting Voices: Dilemmas of Gender, Political Philosophy, and Policy*, edited by Iris Marion Young, 134–164. Princeton: Princeton University Press, 1997.

Index

P. Crawford et al., *Florence Nightingale at Home*,
https://doi.org/10.1007/978-3-030-46534-6

D

E

F